"This is an evolutionarily sophisticated book manuscript that I found very valuable."

John Alcock, *Emeritus Professor, School of Life Sciences, Arizona State University*

"This book is a refreshing counterpoint to the classical and still fashionable reliance on narrative biographical formulations in clinical psychiatry which endure despite a century of countervailing behavioral neuroscience and genetics evidence. The author manages to entertain while tackling this complex topic."

Albert H.C. Wong, *Professor of Psychiatry, University of Toronto, Canada*

Challenging the Therapeutic Narrative

This volume explores and challenges the assumption that behavioral proclivities and pathologies are directly traceable to experience—an assumption that still widely dominates folk psychology as well as the perspective of many mental health practitioners. This tendency continues despite powerful evidence from the field of behavioral genetics that genetic endowment dwarfs other discrete influences on development and psychopathology when extrinsic conditions are not extreme.

An interdisciplinary collection, the book uses historical, cultural, and clinical perspectives to challenge the longstanding notion of identity as the product of a life-narrative. Although the nativist-empiricist debate has been revivified by recent advances in molecular biology, such ideas date back to the Socratic dialogue on the innate mathematical sense possessed by an illiterate slave. The author takes a philosophical and historical approach in revisiting the writings of select figures from science, medicine, and literature whose insights into the potency of inherited factors in behavior were particularly prescient and ran contrary to the modern declivity toward the self as narrative. The final part of the volume uses historical and clinical perspectives to help illuminate the elusive concept of innateness and highlights important ramifications of the revolution in behavioral genetics.

Seeking to challenge the clinical utility of the therapeutic narrative rather than the importance of experience per se, the book will ultimately appeal to psychiatrists, psychologists, and academics from various disciplines working across the fields of behavioral genetics, evolutionary biology, philosophy of science, and the history of science.

Robert G. Goldstein is Clinical Instructor in Psychiatry and Assistant Attending Psychiatrist at Weill Cornell Medical College and New York-Presbyterian Hospital in New York City. He is also a member of the Research Faculty at the DeWitt Wallace Institute of Psychiatry, an interdisciplinary research division at Weill Cornell, USA. He is a graduate of Brown University and the Icahn School of Medicine at Mount Sinai, USA.

Explorations in Mental Health

Trauma and its Impacts on Temporal Experience
New Perspectives from Phenomenology and Psychoanalysis
Selene Mezzalira

Self and Identity
An Exploration of the Development, Constitution and Breakdown
of Human Selfhood
Matthew Tieu

Co-Production in Mental Health
Implementing Policy into Practice
Michael Norton

Towards a Transtheoretical Definition of Countertransference
Re-visioning the Clinician's Intersubjective Experience
Rudy Roman

Challenging the Therapeutic Narrative
Historical and Clinical Perspectives on the Genetics of Behavior
Robert G. Goldstein

Critical Resilience and Thriving in Response to Systemic Oppression
Insights to Inform Social Justice in Critical Times
Melissa L. Morgan

For more information about this series, please visit www.routledge.com/
Explorations-in-Mental-Health/book-series/EXMH

Challenging the Therapeutic Narrative

Historical and Clinical Perspectives on the Genetics of Behavior

Robert G. Goldstein

Routledge
Taylor & Francis Group

NEW YORK AND LONDON

First published 2023
by Routledge
605 Third Avenue, New York, NY 10158

and by Routledge
4 Park Square, Milton Park, Abingdon, Oxon, OX14 4RN

Routledge is an imprint of the Taylor & Francis Group, an informa business

ISBN: 978-1-032-39580-7 (hbk)
ISBN: 978-1-032-39874-7 (pbk)
ISBN: 978-1-003-35179-5 (ebk)

DOI: 10.4324/b23263

Typeset in Times New Roman
by SPi Technologies India Pvt Ltd (Straive)

For Anna, Ethan, and Donna

Contents

Acknowledgments

Thanks are due to a number of people who supported me throughout the writing of this book. John Krasnoff, with his training in behaviorism, taught me much about how to view the patterns of our choices stripped bare of human meanings, values, and strivings. Daniel Kalb tried his best many years ago, in a cheap hotel room in southern Mexico, to introduce me to the credo—follow ethology! I was too young to appreciate his wisdom. Ricky Porter provided sage advice during the writing of this book and dating back to our hallowed times together on Charter Road. Joanna Scheier, a colleague and friend from training days, believed in the project and always lent a sympathetic ear. My mother, always in my corner, bought me an ant farm before I knew what entomology was; my late father bequeathed to me, by way of temperament, a vigorous skepticism. I am grateful, as well, to my sister Ellen who originally inspired my love of nineteenth-century fiction. The renowned expert in animal behavior, Professor John Alcock, generously offered to read the entire manuscript after I wrote to him asking some questions concerning behavioral ecology. Frank Rubino and Barbara O'Dare read snippets of proposals and chapters along the way and lent much appreciated support. My colleagues at the history of psychiatry research division at Weill Cornell Medical College—The DeWitt Wallace Institute for Psychiatry—also provided encouragement and an audience for earlier versions of some of these chapters. Finally, I would like to thank my editor at Routledge, Alice Salt, who saw value in my somewhat unorthodox perspective and skillfully shepherded my manuscript through the process.

Author's Note

Introduction

The field of behavioral genetics has provided powerful evidence that genetic endowment is the largest single influence on an individual's enduring patterns of behavior and psychopathology. Research on twins has also suggested that traditional notions of environmental influence, when not extreme, may have far less impact than previously believed. Despite this, constructing narrative, biographically based explanations of behaviorial proclivities and pathologies is still a remarkably common feature of mental health treatment.

This collection of essays argues against the assumption that behavioral proclivities and pathologies are, in any meaningful way, directly traceable to experience—an assumption that still dominates folk psychology as well as the perspective of many mental health practitioners. This is not to say that experience or environment has scant influence on development or psychopathology; rather, as a number of prominent behavioral geneticists have maintained, these external influences tend to operate in unsystematic ways—yielding incremental, unpredictable effects that are hopelessly entangled with an individual's temperament (Plomin, 2018; Turkheimer, 2000). Because of this, attempts on the part of clinicians to link traits and pathologies to discrete events, relationships, or extrinsic conditions are often highly speculative endeavors.

Furthermore, biographical accounts of behavior are often marked by problematic logic—presupposing that mental symptoms fulfill some sort of goal or purpose. Thus, such accounts often fall prey to a sort of commonsense reasoning that often does not provide worthwhile insight into mental phenomena and their antecedents. This is not to say that the laity or mental health community do not recognize the importance of the revolution in molecular biology: genes are all the rage. And yet, this revolution has had far too little impact on the deeply ingrained tendency to mine our biographies for clues to the contours of our mental lives.

DOI: 10.4324/b23263-1

This volume does not attempt to survey the very large scientific literature on behavioral genetics or molecular biology relevant to human behavior; instead this interdisciplinary collection seeks to provide a novel slant on historical, cultural, and clinical aspects of the nativist/empiricist debate—a debate that has been revivified by advances in molecular biology but which dates back to the Socratic dialogue on the innate mathematical sense possessed by an illiterate slave.

The first chapter of the book, entitled "The Problem with Storytelling," challenges what I refer to as the narrative account of behavior. Although biological psychiatry has garnered significant cultural, academic, and commercial influence, the idea that mental life is structured according to discursive, commonsense rules has stubbornly persisted—not only in treatment settings but in the larger culture as well. This is chiefly due to two factors which are explored in greater detail in subsequent chapters. First, many clinicians unwittingly continue to channel watered-down aspects of Freud's narrative/biographic formulation of development and psychopathology. Although Freud is long dead and buried in academic psychiatry, significant segments of the psychotherapeutic community continue to stress the formative impact of early experience; only now, blocked libidos and incestuous fantasies have been jettisoned in favor of theories stressing, for example, inadequate attachment during childhood as causes of pathology. Also contributing to the persistence of biographical accounts of mental disorders is the human tendency to seek cause-and-effect relationships among seemingly related variables even when causes may be categorically unrelated to the phenomena in question (i.e., psychological difficulties do not necessarily have psychological causes). The first chapter also includes anecdotes from my clinical training and practice that have fueled my interest in behavioral genetics—for example my experience treating parents of adopted children as well as my clinical work with Hasidic Jews who, as members of a pre-Freudian culture, have always intuitively regarded mental disorders as having potent genetic antecedents.

Chapter 2, "The Rise and Slow Decline of the Therapeutic Narrative," reviews how the narrative conception of behavior arose and why it has been so difficult to eradicate. Features of mammalian cognition are highlighted that contribute to the kinds of flawed thinking that characterize narrative/biographic accounts of behavior. For example, it has been shown experimentally that all animals are predisposed to behave as if correlated events reflect causality even when the correlation is specious or arbitrary. Our species is not only prone to this type of error, we are also capable of elaborating such errors into all manner of confabulation. Such confabulations often presuppose that behavior is internalized through observation or experience and that psychological symptoms fulfill some sort of purpose—functioning like protective mental tropisms. But the use of commonsense reasoning to understand the antecedents of behavior is likely problematic for more obvious and basic reasons: the systems underlying behavior are not structured according to discursive, commonsense rules. This is why introspection is generally not well-suited for understanding most mental phenomena—such as, for example, how the brain's visual processing system assembles a percept.

In an attempt to debunk the type of reasoning that frequently characterizes narrative explanations of behavior, I often strike a polemical tone. But the reader should not assume that I am not mindful of some of the problems that have attended the rise of biological psychiatry. Indeed, this book should not be read as an endorsement of bringing a narrow biological perspective to bear on all the difficulties that patients bring to the consulting room. However, the evidence suggests that the parameters that guide our behavior throughout our lives, and that undergird our maladaptive tendencies, are in greatest measure as a percentage of the total, derived from heritable influences. And, as I argue, even the most intuitive clinician would benefit from understanding what science can teach us about these influences.

A reasonable nativist perspective, although embracing a materialist stance, must also recognize the radical importance of extrinsic factors—including economic, cultural, and familial circumstances—in determining the trajectory of each individual life. However, as is also argued, tracing how these disparate forces operate (and deriving a how-the-leopard-got-its-spots account of how we become who we are) is often a misguided exercise: it is not so much that these forces have no impact; rather they operate in such unsystematic and unpredictable ways as to be fundamentally untraceable. Furthermore, the tendency to hunt for biographical explanations of psychopathology has contributed to a therapeutic culture that is often prone to ascribing parental blame. When one tallies up the various potential influences on development and behavior that are not strictly heritable, these are not insignificant. However, in cases where environmental factors are not extreme, these influences are likely so idiosyncratic and incremental that theories ascribing causation to discrete, identifiable, extrinsic factors should be approached with caution.

The book's second part, entitled "Apostles of Modern Nativism," highlights the careers of two scientists, Francis Galton and Seymour Kety, whose work was critical to the development of the field of behavioral genetics. Galton, a cousin of Darwin, developed the first twin studies in an effort to prove that mental characteristics were as heritable as physical traits. In addition, his innovations in biostatistics were critical to the emergence of the discipline of population genetics. Kety, a mid-twentieth-century American researcher, was responsible for one of the largest and most rigorous studies of the inheritance of a major mental illness. Using Danish adoption records, and the population methods Galton had pioneered decades earlier, Kety was able to show a very significant genetic component to schizophrenia. Despite his accomplishments, Kety faced opposition from some psychiatrists and social scientists who appeared to exhibit an ideological hostility toward genetics that, in some respects, resembled Lysenkoism—the anti-Mendelianism that dominated Soviet science earlier in the twentieth century. This part, oddly enough perhaps, also includes a chapter, "The Novelist as Accidental Nativist," on the novelist Honoré de Balzac. It is argued that the founder of literary realism had strong convictions regarding the inborn and inflexible nature of behavior—convictions that perhaps stemmed from the fact that he realized he had inherited his own addictive, reckless temperament from a wildly impulsive father. Interestingly, on more than one occasion, Balzac's characters proclaim that the addict is someone who is *born*, not made, that what he suffers is worse than cholera, requiring a good physician even before a sympathetic magistrate. Balzac's implicit nativism, one could argue, reveals the contemporary notion of identity as the product of a life-narrative to be a historical anomaly.

The final part of the volume, "Reconnoitering Innateness," uses historical and clinical perspectives to help illuminate various aspects of innateness. The first chapter in this section, "Innateness Wars: A Darwinian Aside," explores tensions between opposing camps in evolutionary biology over the last 100 years in the hope of salvaging the embattled concept of innateness—a concept that a number of philosophers of science believe has outlived its usefulness. Of course any discussion of innateness inevitably leads back to Darwin as the nature versus nurture debate had an instructive precursor in the efforts to understand and refine the boundaries between instinctual and learned behavior in the decades after his death. The work of Michael Ghiselin and E. O. Wilson are discussed to illustrate how evolutionists have tried to grapple with the problem that high intelligence presents to an understanding of innateness in our species. This chapter also attempts to address the following question: How does one discern the outlines of innate proclivities in a species such as ours that is characterized by complex states of awareness and high degrees of learning that would appear to swamp or obscure the outlines of instinctual life? As Ghiselin and Wilson have both argued, Darwin's psychological writings contain the seeds of a biological theory of human social behavior: although a myriad of psychological factors obviously play a role in modulating behavior in *Homo sapiens*, Darwin believed that even quite complex behaviors in higher species are mediated through "sheer force of inheritance" and thus the origin of many behavioral proclivities in humans may be independent of the evolution of significant cognitive complexity.

The chapter entitled, "The Missing 50%: Non-Heritable Sources of Variance," delves further into topics in the nativist/empiricist debate that are most relevant to psychiatry. For example, what might account for the non-heritable influences on behavior (estimated to be approximately 50% of the total for many behavioral traits) given that twin studies have shown that traditional types of environmental influence exert fairly minor effects? It turns out that a significant portion of these non-heritable sources of variance may be attributable to processes that regulate gene expression, such as epigenetic effects. Although the epigenetic modulation of gene expression can indeed be influenced by environmental factors, a significant portion of developmental variance appears to be stochastic in nature. Because these sources of variance involve highly random, indeterminate processes, it is easy to see how these reduce concordance rates in monozygotic twins and so inflate the "non-heritable" side of the tally—even though many of these processes may involve little or no systematic environmental influences. Given the dynamic nature of development—a process involving scores of interacting, reciprocal processes—many of the non-heritable (presumed environmental) influences that are not random will likely remain extremely difficult to qualify and so may have a limited place in generalizable theories of human behavior.

This chapter also examines the nativist perspective on those influences on behavior that are indeed biographical. As noted above, even the most ardent nativist must recognize the impact of experience on development. However, the nativist tends to view these extrinsic influences as *releasers* rather than sculptors of behavioral proclivities. Clinical examples are provided to illustrate the paradoxical aspect of development: it is both sensitive to perturbation to a degree, yet highly buffered, and constrained by delimited, pre-specified, developmental alternatives. This chapter also covers other intriguing aspects of the inheritance of mental characteristics: for example, given that most behavioral traits are abstractions (i.e., amalgams of more basic functionalities) and that these traits are likely influenced by scores of genes (not to mention epigenetic as well as gene x gene and gene x environment interactions), how is it possible that there are such a limited number of pathological variants—and that these variants pass so reliably through the generations?

The final chapter, "The Dimensional Approach," addresses how advances in behavioral genetics may eventually allow psychiatric researchers to dissect the behavioral phenotype along more credible dimensions. Understanding behavior in terms of a palette of heritable traits can not only improve diagnostic validity, it may also help clinicians and patients appreciate the relationship among associated traits. The dimensional approach is fundamentally a descriptive, taxonomic approach—one that often eschews questions of ultimate causation. So then, a treatment based on the dimensional perspective opts for trading "why" questions for "how" questions: namely, how do patterns of choice operate under variable conditions? How can identifying enduring proclivities help one utilize reason and self-control to interrupt default patterns of choice and behavior?

Finally, a note on terminology is in order. Throughout the book, I needed a way to allude to two opposing camps in the historical nature/nurture debate. This debate has, of course, been recast as it is universally accepted that almost all traits and disorders emerge from the complex, dynamic interplay of genetic, extrinsic, and stochastic factors. Nonetheless significant colloquy persists between those that believe genetic influence is by far the most powerful factor in behavior and those that are more skeptical of findings in behavioral genetics—often psychotherapists or those who study developmental psychology. I have chosen the terms "nativist camp" and "empiricist" or "anti-nativist camp" to describe these two groups. These are terms descended, in part, from seventeen and eighteenth-century philosophy—specifically from the writings of Hume and Locke among others. I use these terms advisedly—freighted with history as they are—with the caveat that both sides now operate on a relative continuum. Nativism, when applied to human behavior, does not intend to imply a belief that the behavior in question is inborn or innate in the strict sense—like the waggle dance of bees—nor that environmental effects are trivial. Similarly, the "anti-nativist" or "empiricist" labels do not mean to imply a *tabula rasa* version of mind without any antecedent biological structure.

References

Plomin, R. (2018). *Blueprint: how DNA makes us who we are*. Cambridge: The MIT Press.

Turkheimer, E. (2000). Three laws of behavior genetic and what they mean. *Current Directions in Psychological Science*, 9, 5, 160–164.

Part I
The Confabulating Species

1 The Problem with Storytelling

American psychiatry has been largely liberated from the patent absurdities of orthodox Freudianism that held sway in the field for much of the mid-twentieth century. However, mental health treatment still too often bears the influence of a narrative-based understanding of behavior and mental infirmity. Implicit to this view is the notion—originally derived from psychoanalysis—that identifiable early experiences and relationships shape behavioral responses exhibited throughout life and that there is therapeutic utility in attempting to trace how these experiences sculpt one's behaviors and pathologies.

However, twin studies conducted over the last 50 years have suggested that traditional notions of environmental influences, if not extreme, may have modest impacts on the development of behavioral traits and psychiatric disorders. Furthermore, extrinsic influences chiefly operate in unsystematic and idiosyncratic ways and so it is nearly impossible to generalize their role in development. Finally, behavioral geneticists have discovered that environmental influences are themselves highly dependent upon and conditioned by the genetic constitution of the individual and therefore extrinsic conditions tend to exert their effects in a fashion that is hopelessly entangled with temperament. Therefore, the environment is inherently interactive in highly dynamic, reciprocal ways (Turkheimer and Waldron, 2000, 92).

And yet despite this, the faith in the value of a therapeutic narrative has stubbornly persisted and it continues to influence not only psychological treatment but the larger culture as well. The effects of this influence can be seen in countless middlebrow memoirs and biographies, but it is also evident in sophisticated cultural fare. *The Sopranos*, for example, was a finely written drama filled with subtle character portrayals. However, the mob boss's psychiatrist often offered up biographically based theories ascribing many of the mobster's emotional struggles to

DOI: 10.4324/b23263-3

a difficult relationship with his repugnant mother. Similarly, one can find all sorts of speculations by pundits both within and outside the mental health field concerning the biographical origins of psychopathology in public figures—whether in political leaders, notorious criminals, or Hollywood royalty just out of rehab. A significant segment of the public appears to subscribe to such explanatory origin stories—especially as these tales appear to follow the contours of commonsense and are often presented by professionals with advanced degrees and an air of conviction. But leaving aside issues of causation, even if we allow that the mind gains certain aspects of its structure through experience, introspection (undertaken alone or with a therapist) likely yields little about how this structure is acquired or how it might operate. Nonetheless, guided introspection—with the goal of formulating etiological, biographical accounts of behavioral patterns—still constitutes a significant part of psychological treatment in many settings.

I have heard numerous accounts from patients over the years of psychotherapists offering up narrative explanations of psychopathology. Although a good number of clinicians have adopted more pragmatic approaches—relying on cognitive behavioral techniques and the identification of maladaptive behavioral patterns linked to a person's temperamental vulnerabilities—many clinicians still assume that most psychological problems must derive to a significant degree from psychological causes; accordingly, they often seek historical/experiential explanations for their patients' emotional difficulties. Although few of these therapists have had formal psychoanalytic training, they are unwittingly channeling aspects of such approaches.

Narrative explications of psychopathology often display characteristic errors. For example there are etiological conjectures that assume psychological symptoms serve a function or fulfill a purpose—for example, interpersonal avoidance might be thought to result from historical situations or relationships that necessitated self-protection. Another pitfall is the tendency to elevate discrete events or relationships into a sort of developmental lathe that is capable of forging persistent patterns of behavior. This contradicts evidence from behavioral genetics suggesting that transient, discrete events tend to have modest effects on a person's enduring patterns of behavior. Of course truly traumatic experiences can have lasting impacts, but even these often function as releasers rather than sculptors of behavior—triggering latent, genetically based vulnerabilities. In other words, a severe stressor might worsen a pre-existing tendency toward depression but it would be unlikely, for example, to create a persistent proclivity toward social avoidance. One of the more common sources of error in psychological explanations of behavior is the confounding of genetic and environmental influence—a consequence of the fact that most people are raised by those who have also passed their genes down to them. A final type of error ascribes causation to events or experiences that appear thematically related to the pathology or symptom in question. Such explanations are perhaps common because they make humansense; namely they appear to carry a degree of internal logic and coherence. What follows is a clinical example of this type of conflation wherein a therapist ascribes causation to environmental factors based on thematic proximity or congruence.

This clinical example involves a man who was raised from an early age by supportive adoptive parents as his unmarried biological mother struggled with addiction. Despite being happy in a long-term relationship, he had always balked at the thought of marriage and this distressed his live-in girlfriend. Although eschewing substance use, he struggled with online gambling, excessive shopping, and had a garage filled with half-opened purchases. His girlfriend pushed him into psychotherapy and the therapist made a commonsense assumption that his avoidance of marriage, and his escapist addictive behaviors, must be tied to his abandonment by his biological parents (i.e., his hesitance to marry likely reflected a feared reactivation of negative feelings surrounding family life and a need to devalue familial attachments). Despite the apparent hermeneutical ripeness of this theory, this explanation did not resonate with him: he was very close to his adoptive parents; he had never been particularly focused on his biological parents and did not even know his biological mother's whereabouts. He came to regard his aversion to marriage simply as part of a general, life-long pattern of avoidant behavior and procrastination. Through the machinations of social media, however, he eventually obtained more information about his biological mother from other family members. She apparently showed a similar pattern of resistance to long-term commitments to partners as well as excessive, senseless shopping and hoarding of purchases. He was amused by this discovery as it confirmed his intuition that many of these behaviors were base properties without significant psychological antecedents. These behaviors were, it turned out, simply part of his mother's legacy—like his garage full of useless purchases. Now such a genetically based theory of the origins of his pattern of delay and avoidance does not imply that his broken home had no psychological impact on him. However, just because behaviors or symptoms appear thematically related to earlier life experiences does not necessarily suggest they are causally related.

In fact, the assumption that correlation or congruence between variables implies causation is such a common type of error that it enjoys the status of a formal logical fallacy: *cum hoc ergo propter hoc* (with this, therefore because of this).

It would seem that we have already stumbled into the tired, tortuous terrain of "nature versus nurture." Suffice it to say, this dichotomy has outlived its usefulness and has been largely reframed: it is now taken for granted that both genetic *and* non-genetic factors interact in a highly dynamic fashion to produce most types of psychopathology. Indeed advances in molecular biology have proven that genetic expression is often modulated in an interactive fashion by extrinsic influences—therefore revealing the speciousness of the nature/nurture duality. And of course it goes without saying that our narrative arc is of utmost import: we can be saved or undone by the course of events. However, no matter what befalls us, the ways in which we respond to life's contingencies are overwhelmingly a function of innate proclivities; and these proclivities remain strikingly consistent throughout our lives. And so, after discounting the influence of luck, environmental factors, when not unusually extreme, are either significantly overshadowed by genetic endowment or else these factors operate in ways that are so difficult to trace that devising biographic, commonsense accounts of enduring behaviorial patterns is often hopelessly speculative. Furthermore, as outlined above, such accounts are often marked by problematic teleological thinking—presupposing that natural phenomena (in this case symptoms or traits) display what biologist Michael Ghiselin, in another context, has called "terminating orientation" (i.e., that they fulfill a goal or purpose), or that they embody "intelligible orderliness" in terms of design and function, and/or that they embody a sort of "transcendent rationality" (Ghiselin, 1974, 21).

Although my nativist perspective has been influenced by empirical findings in the field of behavioral genetics, my clinical experience as a psychiatrist has also shaped my views. I have had the opportunity to observe patterns of inheritance in my patients' families, but most compelling have been the accounts of patients who have adopted children. These scientifically naïve patients frequently come to realize that their adopted children display behavioral difficulties that are completely distinct from the problems that exist in themselves or their genetically related kin. Because of this, many of them find it particularly difficult to understand their adopted children's behavior. Despite their sacrifice and devotion, these parents frequently report being subject to blame by therapists who attempt to find environmental explanations for their children's difficulties. They are often relieved to hear me explain that their children are more likely simply exhibiting the pathologies that are adrift in their families of origin. These parents no doubt fell short, as all parents do in certain respects, but in most cases they had provided supportive environments. This likely improved their children's long-term prospects to a degree but could not alter their fundamental proclivities.

Twin studies carried out over the last 40–50 years support this view. Such studies have provided compelling evidence that traditional notions of "nurture" (when not extreme in its variance) contribute a small proportion of systematic influences on behavioral traits and psychopathology: for most psychological traits, monozygotic twins raised apart show a concordance (meaning the proportion that share the trait in question) of 40–50%, but remarkably this level of concordance, for most traits and conditions, is not appreciably higher in identical twins that are raised together. This suggests that the influence of the so-called "shared environment" is often quite modest. Further confirmation of this comes from studies of unrelated adopted children ("adoptive siblings"). For most behavioral traits, concordance rates in these non-kin siblings raised in the same home approached those found in random strangers (Plonim, 1987). These studies do not prove that environment has no influence—on the contrary concordances of 40–50% in identical twins raised together implies a significant role for some sort of non-heritable influence. However, these studies suggest that traditional notions of nurture—one's home environment that is shared with one's siblings—likely contribute modest effects. Because of this finding, behavioral geneticists originally invoked a concept they called the "non-shared environment" (extrinsic factors unique to each sibling raised in the same family) to account for the non-trivial behavioral divergence found in monozygotic twins. However, important non-shared environmental influences turned out to be difficult to qualify and those that did rise to significance in some studies tended to exert very small effects (Plomin, 2011). Because of this, the very distinction between shared and non-shared environmental influences has been questioned: it may matter less whether external factors are objectively shared but rather the critical issue is how extrinsic factors interact with each person's unique phenotype and developmental course. Again, discrete environmental variables are profoundly interactive and appear to carry small direct effects.

Because of this, many behavioral and developmental geneticists began to question the utility of trying to study these non-shared environmental influences. Even if some rose to statistical significance they were only responsible for small sources of behavioral variance and were difficult to replicate across populations. This eventually led Plomin to fully embrace what he dubbed "the gloomy prospect": namely, the hunt for systematic non-shared extrinsic influences is likely futile as these factors tend to exert their effects in an indeterminant and transient manner. This contrasts with the genome which by far supplies the largest and most enduring influence on our behavior (Plomin, 2018, 80).

Even those behavioral geneticists who believe Plomin leans too far in the nativist direction, tend to acknowledge this gloomy prospect. But apart from this fact—the forbidding challenges of studying environmental influences on development—it may be misleading to consider many of these non-heritable sources of variance as truly extrinsic. This is so because a significant percentage of these non-heritable factors appear to be related to events operating at the molecular level *within* the developing organism and often involve little to no systematic external influence. These include contingencies that affect how various neural tracts and synaptic connections are organized during fetal development, as well as processes that modulate gene expression—such as epigenetic effects. As I will discuss in Chapter 7, while these non-heritable influences can indeed be impacted by the environment, many of these developmental and epigenetic processes are highly stochastic in nature. In addition there are the often random human-scale events unique to each individual—the mishaps and fortuitous occurrences that impact one's personal course in the idiosyncratic manner Plomin and others have highlighted. Therefore, how development unfolds rarely adheres to any discernible or predictable pattern, certainly not one ruled by commonsense or narrative logic.

Therefore, after being subjected to all sorts of biographical theories about the causes of their children's problems—theories of kids missing their birth mothers, of parental over-involvement or under-involvement, of permissiveness or excessive control—these parents of adopted children usually come to realize that they know better than well-intentioned therapists who had not raised adopted children. Many of them finally accept that their children had arrived with problems and proclivities that were pre-specified according to the lineaments of their ancestral bloodlines. They realize that a biographical explanation of behavior is often a fool's errand.

What was patently obvious in these adoptive families is often less apparent in typical families where genetic and environmental influences are superimposed and so more easily confounded. The impetus for writing this book stemmed, in part, from repeated complaints by my patients that therapists to whom they sent their children often devolved biographical explanations for their children's problems that implicitly or explicitly blamed their parenting for their children's struggles. This sometimes served to alienate these children from their natural caregivers who, despite their foibles and imperfections, were the most important sources of material and emotional support for these children.

The fact that behavior and mental pathologies are influenced to only a limited degree by experience has been recognized since antiquity. According to Robert Burton's Renaissance compendium of mental infirmity, *The Anatomy of Melancholy* (1621), Hippocrates argued that "in manners and conditions of the mind the character of the parents is transmitted to the children through the seed." Burton, himself an apparent nativist, contended

> "such a mother, such a daughter; their very affections … to follow their seed … and the malice and bad conditions of the children are many times wholly to be imputed to their parents; I need not therefore make any doubt of melancholy, but that it is an hereditary disease".
>
> (Burton, 2001, 212)

Shakespeare, one of the first English speakers to employ the nature/nurture opposition, described Caliban as a "devil, a born devil, on whose nature/Nurture can never stick." Perhaps it is no accident that the playwright presages nativist thinking: given the constraints of his craft, the dramatist must create a character within the space of a few hundred lines. Such shorthand rendering is possible because playwriting is, in fact, a process of evocation—the summoning of types already intuitively understood by the audience. Of course each person is unique in certain respects, surely in the particularities of their experiences. However, individuals display traits drawn from a limited palette—limited by the highly specified algorithms that are key to regulating behavior in our species. These traits coalesce into a finite number of arrays that, in turn, constitute a finite variety of natural human kinds.

The modern discipline of psychiatry is often ridiculed for reducing the rich complexity of human behavior to what is, in essence, a taxonomy of pathological types specified by lists of signs and symptoms. The DSM-5, the diagnostic manual used by psychiatrists, is organized into sections, each dealing with general classes of disorders such as anxiety disorders, psychotic disorders, addictive disorders, mood disorders, and personality disorders. The specific diagnoses within these broad categories vary in terms of their cogency: some, like bipolar disorder for example, probably reflect true syndromes while others are far more amorphous—overlapping with other diagnoses or with traits commonly found in well-adapted individuals. The DSM is certainly an imperfect document, occasionally redolent of an actuarial table. But I find the common critique aimed at the DSM—that it is simple-minded, even vulgar in its reductionism—to be partially true but beside the point: all pragmatic forms of knowledge are limited by the necessities of the task they are designed to fulfill. In order to treat patients, physicians must be able to create diagnostic categories—however inexact or provisional. And psychiatry faces a special challenge in that there are no objective data external to the diagnostic criteria themselves to support the diagnoses (such as a blood test or tissue biopsy). As psychiatry lacks so-called "criteria validity," it must rely upon what is known as "construct validity"; namely, for a given theorized diagnosis, validity depends only upon how well the supposed characteristics of the disorder correlate with each other. These characteristics can include not only symptoms but also associated features such as age of onset or the presence of a family history. As Aboraya writes in her review of validity in psychiatric diagnosis: "construct validity boils down to the circumstantial evidence for the *usefulness* of the construct or hypothesis under study" (Aboraya et al., 2005, 50).

By the time I completed medical school in the late 1980s, Prozac (the first in a class of drugs that would revolutionize psychiatry) had received approval from the FDA. However, much of academic psychiatry had yet to fully emerge from its roots in that nineteenth-century amalgam of philology and Helmholtz-inspired energetics known as psychoanalysis. Many of the instructors at my residency program had been trained as analysts and they tended to regard biological approaches to psychopathology as not only reductionist but also anti-humanist in that they neglected the patient's personal life story. While recognizing that severe mental disorders like schizophrenia were likely discrete diseases, they still cleaved to a biographical-based conception of milder mental disturbances and "character disorders."

Psychoanalysis has since lost most of its influence in academic psychiatry while biological approaches have gained ascendence. Nonetheless, as I argued above, a version of psychoanalytically tinged thinking has stubbornly hung on—having so thoroughly infected Western culture and folk psychology. But perhaps it has hung on not due to its intrinsic epistemic heft but rather because it is consonant with a general human tendency to read coherence in seemingly related variables (more on this to follow in the next chapter). Many people, including many therapists, still make the commonsense assumption that a person's behavioral problems and proclivities are significantly shaped by one's close relationships and emotionally laden experiences. In therapists' offices throughout the world, a watered-down version of Freud's approach often plays some role in the therapeutic process as the patient (with the therapist's guidance) devises a biographic schematic of his or her current difficulties. But these narrative expositions of behavior are often based on the fallacy that "like begets like": negative emotional states in the present must have originated from prior negative emotional experiences. Such fallacies were common in pre-scientific medicine: malaria, a condition associated with humid, warm climes, must surely be caused by humid, warm vapors. The gap left by the yet undiscovered malaria parasite was filled in with a commonsense confabulation.

As I will discuss in greater detail in the following chapter, confabulation may be one of the essential activities of human cognition; indeed, the neural mechanisms that ascribe cause-and-effect relationships appear to hum along in the background whether the brain has veridical data handy or not. The gifted Dr. Freud had a prodigal confabulatory apparatus: he confabulated an entire branch of therapeutics that operated under the assumption that behavior is structured along narrative rules. But in actuality, our enduring proclivities are highly determined by a ternary code—the base pairs that constitute our genetic endowment. Of course in Freud's day, the mechanics of heredity were yet to be laid bare. But that, it turns out, is a poor excuse. As I will discuss in Chapter 3, a contemporary of Freud's, Francis Galton, had no access to a credible science of heredity yet he was convinced that mental infirmity, personality, and various talents were predominantly a function of inheritance. Though Galton can be considered the father of behavioral genetics, the real hero of our story is his cousin, Charles Darwin; it was the Darwinian revolution that inspired Galton to undertake the scientific study of the inheritance of mental characteristics. Darwin himself had argued that behavioral traits be treated no differently from physical ones—both having been shaped by natural selection. He also very much believed, as William James would argue some years later, that as hominids developed increased abstract reasoning abilities, they did not shed their instinctual heritage. On the contrary, the superior intelligence of *Homo sapiens* endowed the species with a greater not diminished repertoire of instinctual behaviors, and variations in these innate proclivities descend through human pedigrees.

Just as the farmer or hunter can appraise the temperament of a draft horse or hound, so too can the skilled salesman or grifter easily recognize behavioral proclivities in people. Each of us may present a unique blend of various dimensions of temperament, however the lathe of natural selection has operated upon a finite number of behavioral repertoires. As Galton pointed out when discussing the innate proclivities of animals suited to domestication, the behavioral pro-clivities of all species are variegated but also essentially limited: ask anyone who has tried to domesticate a zebra.

No one, not even an arch nativist, could sensibly argue that the con-texts within which our natures are expressed—be they familial, social, or cultural—are trivial in their impact. These influences often impact the occupation we choose, our table manners, whom we marry, the economic status we achieve. And, of course, environmental stressors (whether due to poor nurturance or bad luck) often play a significant role in bringing latent vulnerabilities to the fore. However, leaving aside extreme extrinsic circumstances, the single largest influence by far on one's behavioral proclivities and mental pathologies has been proved (through decades of painstaking research) to be one's genetic endow-ment. As previously noted, the concordance rates for most behavioral traits in monozygotic twins adopted away at birth fall within the range of 40–50%. As twin adoption studies control for environment, and monozygotic twins (MZTs) share 100% of their DNA sequences, con-cordance rates in twins are one method of measuring what geneticists refer to as heritability. Heritability is defined as the degree of variance in a trait in a given population that can be attributed to genetic differ-ences. Now heritabilities of approximately 50% may not sound over-whelming—and some mistakenly conclude that environmental influences account for the full remaining 50% (more on this in Chapter 7). However, as behavioral geneticist Robert Plomin has argued, a 50% effect size is stunning in terms of the amount of variance genetic dif-ferences can explain as few influences in psychology show effect sizes that exceed the low single digits (Plomin, 2018, 31).

Apart from under-appreciating the magnitude of the genetic influ-ence on psychological traits, many developmental psychologists also frequently decry the genetic determinism they believe is implicit to the project of trying to tie complex behavior to genetic endowment. However, no competent behavioral geneticist or molecular biologist believes that heritability exists in a vacuum; on the contrary, it is a statistical measure based on variance in a trait *in a given population*—and a population exists under specific conditions. Thus the heritability of a trait is not a fixed value but will vary based upon variable extrinsic factors. This is consistent with the commonsense conclusion that genetic effects can be amplified or diminished under varying condi-tions—particularly under extreme conditions. A simple example can elucidate this principle. The heritability of height in most populations is quite high—approximately 80%. However, if a population of chil-dren were raised under extreme conditions (such as prolonged famine), many would fall off their predicted growth curves and so the heritability of adult height would likely drop significantly. As will be discussed in later chapters, the field of behavioral genetics has actually helped elucidate the dynamic, interactive relationship between genetic and environmental influences.

As argued above, the influence of Freud on mid-twentieth-century psychiatry played a unique role in obscuring the central role of inheritance in mental illness. However, most scientifically naïve cultures intuited that psychiatric disorders, like other infirmities, ran in families. This may explain, at least in part, why many traditional cultures historically favored the custom of arranged marriage—representing, as it does, the assertion of parental prerogative over the destiny of the bloodline. During my residency training, I had the opportunity to care for a number of Hasidic Jews and I observed their explicit concerns regarding how familial mental illness might influence the marriage prospects of their children or siblings. These concerns were reflected in an extreme wariness of seeking psychiatric services. My old-school supervisors were convinced that this avoidance stemmed from *animus* toward the apostate Freud—a sex-obsessed atheist. However, this insular group, many of whom spoke Yiddish as a first language, was in actuality a pre-Freudian culture—frozen in nineteenth-century *shtetl* life; most of them could not pick Freud out of a police lineup. Nor were they squeamish about discussing personal matters, even sex, with either doctors or religious authorities. However, one thing was certain—they did not want to be seen within 100 miles of a psychiatric facility. The reason for this, I soon realized, was quite obvious: if they or any family member were known to suffer from mental illness, the marriage prospects of all unmarried family members would be compromised. The Freudian revolution had completely bypassed these people; they never doubted that mental problems were inborn and that no one with reasonable alternatives would marry into families plagued by such scourges.

I continued to treat a few Hasidic patients once in practice and I saw first hand that courtship in this community often entailed more negotiation than romance. Some parents of patients even tried to draw me into matchmaking. On one such occasion, the mother of a young patient with a severe disorder called me for advice concerning various suitors being offered up to her daughter. As this patient's condition was known within the community, the family was forced to choose among various men with defects of corresponding degree: "What is worse Dr., a boy with fits (epilepsy) or one with four fingers on one hand?" These were, of course, questions beyond my ken. "Ask the Rabbi," was my customary dodge.

Beginning in the early 1970s, psychiatry began to gradually shed its mid-century Freudian roots and begin its gradual evolution into an empirically based discipline. The epicenter of this change came not from the traditional bastions of academic power on the East Coast but rather from the Midwest. Fifty years ago, at Washington University in St. Louis, Drs. Samuel Guze, Eli Robins, and George Winokur initiated an assault on the psychoanalytic establishment. These clinician-researchers believed that psychiatric disorders, like other medical disorders, were in large measure biological. Thus, for psychiatry to become a bona fide medical discipline, the field had to develop categorical diagnoses with proven reliability and validity. These pioneers from Washington University embraced their role as iconoclasts: they recruited residents with research experience in basic science and apparently hung a portrait of Freud in the department's lavatory. Nor were they squeamish about funding sources: after the risks of smoking became public, the tobacco industry was keen to study the neuropharmacology of nicotine—given its apparent salutary effects on attention and anxiety. The industry, therefore, became one of the underwriters of early biological psychiatry.

The diagnostic schema that Dr. Guze and colleagues helped pioneer (now memorialized in the DSM-5) has been sharply criticized over the years—especially by psychoanalysts and by academics in the humanities. Nonetheless, the categories applicable to many of the major disorders (such as the major mood and psychotic disorders) have held up rather well. Although some of these are in need of refinement (and objective correlates are still sorely lacking), clinicians can usually recognize and agree on these diagnoses. Furthermore, patients who share a diagnosis generally show predictable responses to specific classes of drugs. However, one major group of disorders, the so-called personality disorders (e.g., narcissistic, borderline, anti-social to name a few), has proved more challenging to validate and remains more controversial. This should come as no surprise as, by their very nature, personality disorders encompass a broad array of proclivities that cut across many domains of mental function and influence almost every aspect of a person's life. Thus, in dealing with personality disturbances, the developers of the DSM were faced with the added challenge of deciding which constellations of behavioral traits corresponded to credible types—that is, types that exhibited adequate internal consistency (construct validity).

With the revolution in molecular biology, geneticists garnered the tools to potentially begin linking psychiatric disorders to specific neural mechanisms and genes. However, the challenge has proved daunting: psychiatric disorders are highly heterogeneous (there are likely a number of subtypes of each disorder); further complicating matters is the fact that behavioral disorders are highly polygenic (involve scores of interacting genes). If the task appears complex with regard to seemingly discrete disorders, like obsessive compulsive disorder or schizophrenia, then investigating the genetics of personality—for reasons just discussed—would appear almost impossible. Therefore, in order to attempt to study the genetics of personality, behavioral geneticists realized they must first parse the mental phenotype into more fine-grained dimensions. So instead of looking, for example, for genes associated with antisocial personality disorder, researchers might try to identify genes linked to a specific biological marker—such as diminished fear responses to threatening or disturbing stimuli. One geneticist who, early on, championed a dimensional approach is Robert Cloninger. Based on adoption studies of twins, Cloninger distinguished four dimensions of temperament that he argued showed independent inheritance: harm avoidance (characterized by anxiety and risk-aversion), novelty seeking (marked by impulsivity and low frustration tolerance), reward dependence (found in persons who are approval-seeking and highly sensitive to interpersonal feedback), and persistence (marked by resilience and tenacity) (Cloninger, 1987). Another pioneer of a dimensional approach to personality was the psychiatrist Larry Siever. He advocated the study of intermediate or bridging traits that might constitute credible biological substrates of the personality disorders. These intermediate traits, such as the impulsive aggression or affect dysregulation found in borderline personality disorder, could then hopefully be linked to measurable neurophysiological features which, in turn, might be linked to specific neural and transmitter systems (Siever, 2005).

Although, for the pragmatic purposes of diagnosis, psychiatry divides the behavioral phenotype into discrete disorders defined by specific signs and symptoms, the reader should bear in mind that, in the real world, firm boundaries separating the various maladies seldom exist; furthermore the various domains of behavior often interact. For example, if a person has a personality type marked by significant interpersonal sensitivity, this trait may be intensified by a depressive episode. I tend, in this volume, to focus more on behavioral traits and less on the various disorders such as bipolar disorder or obsessive compulsive disorder. Although these disorders show significant heritability, I chose to give them short shrift as most readers already intuitively regard such conditions as fairly well-defined disorders with significant genetic antecedents.

Nativism has traditionally drawn the ire of many in both scientific and intellectual circles. As I will discuss in Chapter 3, behavioral genetics as a discipline was tainted from its inception by the role of its founder, Francis Galton, in the eugenics movement. Although such criticisms are no longer relevant, there are still many who express concern, for a variety of reasons, about psychiatry's increasing focus on the biological and genetic basis of behavior. These attacks are usually based upon three areas of concern: first, the charge of simple-minded genetic determinism and the neglect of socio-cultural influences on behavior; second, epistemic and scientific concerns (related to what we can hope to know about things as seemingly fuzzy and multi-determined as emotions); and last, concerns about the ethical implications of emphasizing the genetic determinants of behavior. Each of these topics is worthy of a shelf full of books; nonetheless, I will offer some cursory comments on each of the above, starting in reverse order.

Critics of the nativist camp often express the concern that emphasizing the genetic underpinnings of behavior erodes notions of free will and personal responsibility. They argue, for example, that research highlighting genetic commonalities between drug addiction and so-called behavioral addictions, like excessive sexual behavior, downplays the role of personal agency and implies that these behaviors, once primed, carry a degree of inevitability in the genetically vulnerable. However, honest investigators of the biology of addiction should not be held accountable for normative or legal appropriations (or misappropriations) of science. Besides, as a society we may still choose to hold people accountable for their behavior even if we believe we would behave similarly if born with the same genetic liabilities. Indeed our society has shown no qualms about incarcerating millions of people whose addictions and/or mental illnesses have led them to break the law. I cannot imagine how scientifically informed understandings of various forms of addiction and impulse dyscontrol could make matters any worse. Ironically, as the reader will see in my discussion of Balzac's depiction of sex addiction, even the church-going bourgeoisie in nineteenth-century France knew a sex addict when they saw one. And remarkably, they viewed him as one afflicted with a mental infirmity.

Turning to epistemic concerns, for many decades neuroscience tended to neglect the study of emotions: these were regarded as ill-defined, mercurial, too difficult to study in the laboratory (Damasio, 2000). Researchers in cognitive science were only too happy to leave such matters to clinical psychiatry. Meanwhile behaviorists, such as B. F. Skinner, regarded emotions as those fictional causes that people commonly mistake to be determinants of behavior (Panksepp, 1998, 9). As researchers began to develop methods of qualifying and studying emotion, critics—including many clinicians—cried foul: how could neuroscience or genetics ever help untangle the knot of a failing marriage or motivate an addict to stay clean? There are aspects of human experience, such critics argue, that will always be transacted apart from the levels of explanation proper to neuroscience. But it would seem that many have, excuse the tautology, a sentimental attachment to our emotional lives. We are willing to believe that natural selection has produced highly specific, innate capacities for purely "cognitive" functions like visual-spatial or linguistic processing. However, when it comes to emotion, some wish to build a fence around *psyche* and protect her from white-coated philistines. But the emotions—like other critical regulatory mechanisms—have been shaped by natural selection and so are underpinned by substrates that are as discretely specified as any other critical regulatory mechanism. And of course, one would not expect a functional MRI scan to help save a troubled marriage any more than one would visit a stem cell laboratory to manage one's diabetes medications. Nonetheless, one should never consult a marriage or addiction counselor whose understanding of behavior is not informed with what science can teach us about behavior.

The social science critique of nativism often hinges on the charge of genetic determinism. But as evolutionary biologists and geneticists are quick to point out, it is a given that behavioral traits tend to exhibit high levels of phenotypic plasticity—that is, the expression of genes associated with most behaviors is highly context-dependent and influenced by environmental factors. Accordingly, although a certain behavioral trait may show high levels of heritability, there are rarely deterministic genes *for* a behavior or trait; rather there are genes that specify architectural and/or functional variations in nervous systems that are associated with the emergence of a given trait under certain conditions. That said, the sensible nativist does indeed view environmental factors differently than most in the social sciences. The nativist would view most environmental inputs less as *sculptors* or shapers of behavioral proclivities and more as *releasers*—vectors that toggle the organism down fairly restricted, alternative developmental pathways. Culture-bound syndromes, like anorexia nervosa, are good examples of how behaviors can be both responsive to variations in cultural and social conditions and yet quite constrained within the bounds of delimited developmental alternatives. There is, of course, little doubt about the role of cultural norms and pressures on the development of eating disorders in young women raised in industrialized societies. However, although the explicit form the disorder takes is clearly culturally determined, the proclivities that form the substrate out of which this disorder emerges are highly heritable, as well as fairly static and fixed. These proclivities—including perfectionism and obsessive-ness—would, in the absence of cultural factors abetting anorexia, still be passed down the generations. However, in another cultural context, these traits might manifest in a different manner—for example, in the obsessive pursuit of spiritual purity or religious ideals. Thus the innate behavioral syndrome may be conceptualized as a latent susceptibility that may assume different characteristics under varying social, histori-cal, or cultural conditions.

But such scholastic debates concerning determinism are downright civil compared with the ideologically driven and overtly politicized controversies that have stalked the field of behavioral genetics in the 100 years since its inception. This is not surprising given that many of the field's earliest researchers were also founders of the eugenics move-ment—a movement that was later embraced by the Nazis. A stain of this magnitude is difficult to erase—especially as some on the political right continue to misinterpret and appropriate findings in the field for ideological purposes. But as Kathryn Harden laments in her recent book, *The Genetic Lottery: Why DNA Matters for Social Equality*, being a self-described progressive has not immunized her from attacks from the far left. At least one colleague lambasted her because of her interest in studying genetic factors that might influence educational and economic outcomes (Harden, 2021). As a psychiatrist, I am less interested in the study of IQ or educational attainment and more engaged with research on psychiatric disorders. Nonetheless, I too have been met with occasional umbrage from psychotherapists—less for overt political reasons but rather because they mistake my enthusi-asm for genetics for a lack of interest in patients' lived experiences, or worse, as evidence of some intrinsic anti-humanism or therapeutic nihilism. Ironically, a reasonable nativist perspective may be the best corrective to the far-right libertarian view that ascending the social ladder requires only sufficient virtue and will-power. It can also serve as a brake on various forms of discrimination; indeed it could be argued that obesity, homosexuality, and addiction have been de-stigmatized, in part, due to our appreciation of the genetic underpinnings of human differences.

So why, despite the triumphs and prestige of genetic medical research, does this corner of research still struggle with intractable colloquy? As I discuss at greater length in Chapter 6, both behavioral genetics (and evolutionary psychology) are inextricably linked to controversy because they touch upon some of the thorniest human problems: Who or what is to blame for our unhappiness? Can morality be inculcated in children or is it, like any other behavioral trait, significantly influenced by our genes? Can we alter our enduring patterns of choice and behavior or are they relatively static and resistant to change?

By focusing on cultural, historical, and clinical aspects of these debates, I do not intend to gloss over legitimate scientific objections to certain findings put forward by behavioral genetics researchers. To be sure, research in behavioral genetics has had its share of sloppiness and overreach. Such issues will be discussed in greater detail in Chapters 6 and 7. But some of the enduring controversy may stem less from fundamental scientific differences, but rather from the limitations inherent to the study of processes (inheritance and development) that are so complex and opaque.

Perhaps then, the extreme difficulty of studying the genetics of human behavior makes it hard for even behavioral geneticists to play nicely together. As will be discussed in Chapter 7, two leading researchers in the field, Eric Turkheimer and Robert Plomin, appear to have fallen into some measure of discord as Turkheimer seems to believe that his colleague has mistaken the *difficulty* of identifying environmental influences on behavioral traits for the relative unimportance of these influences—at least in comparison with genetic effects. Nonetheless all now agree that, given what we know about the modulation of genetic expression, nature *and* nurture are obviously co-mingled in a developmental process that is dizzyingly interactive, dynamic, and indeterminant. Thus Turkheimer and Plomin both subscribe to the above mentioned "gloomy prospect": the extreme difficulty of identifying environmental influences that are systematic, generalizable across individuals, or that can account for more than a few percentage points of variance. But one could argue, and perhaps Turkheimer and Plomin would agree, that although nature/nurture discussions have been recast they need not be abandoned: how much and in what manner a trait or disorder is influenced by genetic and extrinsic factors remain valid and important research goals—despite the steep practical and theoretical impediments. In some domains, where there are high degrees of gene x environment interaction and where traits are at a greater remove from more modular functionalities, efforts at teasing apart the effects of nature and nurture may prove pointless. However, for traits that display less plasticity, where heritabilities are very high, and/or where we have some understanding of the molecular biology involved, attempts to qualify genetic, epigenetic, and extrinsic influences and mechanisms can yield important results. In the meanwhile, those who lean in the nativist direction will continue to attract criticism—due, in part, to overreach by some—but also because the nativist view implies a pessimistic view of human agency.

But whether or not the difficulty of identifying stable, systematic, extrinsic influences reflects the limits of current science or the relative paucity of such influences, the consequences for the argument of this book would not change: at this point in time, narrative or biographically based explanations of behavior and pathology have little empirical support. Again, this does not mean that environmental factors are unimportant: economic, social, even familial conditions can make or break a person; gifted, intuitive therapists help many people; of course extreme environments can severely impact people. However, because extrinsic influences tend to operate in such idiosyncratic and incremental ways, generalizable theories of biographical causation in psychology remain highly speculative. Furthermore, those that lean toward the nativist camp believe that extrinsic factors most often operate as releasors or vectors. Such vectors toggle the organism among fairly fixed and delimited developmental alternatives rather than functioning as sculptors of proclivities, pathologies, and identities.

Finally, because I anticipate that aspects of my polemic will be misinterpreted, an additional caveat is perhaps in order. As I have emphasized above, I do indeed believe environmental contingencies matter, and yet I am skeptical of most biographical accounts of behavior and pathology. I am, of course, as much a creature of narrative—a creator of meanings to live by—as the next person. What I am attempting to challenge is not the importance of experience per se but rather the clinical utility of the therapeutic narrative. In an effort to establish a veridical schematic of maladaption, the therapeutic narrative attempts to wedge our chaotic real-life experiences into a spurious, deterministic logic. According to this logic, suffering has discernible origins as well as a certain instrumentality. Natural phenomena, however, are far more elusive than this—operating as they do according to rules that often have little to do with human sense or the moralism of fables.

References

Aboraya, A., France, C., Young, J. Curci, K., LePage, J. (2005). The validity of psychiatric diagnosis revisited: the clinician's guide to improve the validity of psychiatric diagnosis. *Psychiatry*, September, 48–55.

Burton, R. (2001). *The anatomy of melancholy*. New York: The New York Review of Books.

Cloninger, C. R., (1987). A systematic method for clinical description and classification of personality variants: a proposal. *Archives of General Psychiatry*, 44(6), 573–588.

Damasio, A., (2000). A second chance for emotion. In Lane, R., Nadel, L., Allen, J., Kaszniak, A. (Eds), *Cognitive neuroscience of emotion*. Oxford: Oxford University Press.

Ghiselin, M. (1974). *The economy of nature and the evolution of sex*. Berkeley: University of California Press.

Harden, K. P. (2021). *The genetic lottery: why DNA matters for social equality*. Princeton: Princeton University Press.

Panksepp, J. (1998). *Affective Neuroscience*. New York: Oxford University Press.

Plomin, R. (2011). Commentary: Why are children in the same family so different? Non-shared environment three decades later. *International Journal of Epidemiology*, 40, 582–592.

Plomin, R. (2018). *Blueprint: how DNA makes us who we are*. Cambridge: The MIT Press.

Plonim, R. (1987). Why are children in the same family so different from one another. *The Behavioral and Brain Sciences*, 10, 1–16.

Siever, L. J. (2005). Endophenotypes in the personality disorders. *Dialogues in Clinical Neuroscience*, 7:2, 139–151.

Turkheimer, E., Waldron, M. (2000). Nonshared environment: a theoretical, methodological and quantitative review. *Psychological Bulletin*, 126, 78–108.

2 The Rise and Slow Decline of the Therapeutic Narrative

Perhaps the simplest and most cogent argument against biographical or narrative-based explanations of psychopathology is that biological systems do not operate according to the discursive logic that governs such explanations. The prior chapter was entitled "The Problem with Storytelling" but it could have as easily been called "The Problem with Commonsense." Commonsense modes of reasoning discern relationships based on temporal, spatial, and categorical features and use these to infer causality. But this default mode of learning and reasoning often yields specious outputs. Consider the ways by which pre-scientific peoples explain natural phenomena, such as crop failure. Such explanations commonly involve supernaturalism but often also exhibit confabulated deductive reasoning. For example, a group believes that the natural world is ruled by thin-skinned gods who keep score and hold grudges; therefore if crops fail, some transgression against a deity must have occurred. Ironically, the Theban myth of Oedipus, which also functions as the founding myth of Freud's pseudoscience, is just this sort of explanation—invoking the crime of incest as an explanation for plague. Humans appear to apply this type of schematic reasoning reflexively—especially when empirically derived knowledge is in short supply.

Why is this tendency so pervasive in human thought? The explanation can perhaps be found in the behavior of lower animals that, in their own way, deploy similar types of cognitive algorithms. Indeed the mental apparatus of all animals appears predisposed to treat correlated events as causally related even when the correlation is specious or arbitrary (Panksepp, 1998, 161). This should come as no surprise as the association of cues in the environment with contingent events (especially threats or rewards) constitutes a basic form of learning with obvious adaptive value; therefore causation-reading may be a default tendency, promiscuously applied. This proclivity is illustrated

DOI: 10.4324/b23263-4

by the avian behavior known as "autoshaping." When pigeons are exposed to an illuminated key just prior to the delivery of food that is provided according to a fixed-interval schedule, they begin to peck at the key when it lights up despite the fact that there is no connection between the animals' pecking and the delivery of the food. If a sentient being were doing the pecking, one would assume that the agent "believed" its pecking behavior had precipitated the food delivery despite the fact that the pecking is purposeless (Panksepp, 1998, 161).

Skinner described a similar type of behavior in pigeons that he explicitly related to the human tendency to attribute causation to temporally correlated events. In this case, Skinner observed that pigeons on a fixed-interval schedule of reinforcement (i.e., when the first response is rewarded after a specified time interval) often behaved as if certain random behaviors they had performed during the inter-reinforcement intervals were causally linked to the delivery of the reward. For example, if a pigeon happened to flap its wings just prior to the delivery of the reward, in subsequent intervals it would flap its wings as if this incidental behavior had meaningful consequences. Never one to avoid going out on a limb, Skinner went so far as to label these "superstitious" behaviors—linking them to certain forms of human magical or wishful thinking such as is displayed by the gambler who always stands in his lucky spot at the craps table. Perhaps Skinner had simply observed the phylogenetic roots of what philosopher David Hume discovered over 200 years ago: causality-reading is an inexorable habit of cognition—spreading itself upon the world rather indiscriminately.

Such linking of random events with desired or feared outcomes may be one wired-in tendency among many that lead *Homo sapiens* to indiscriminately read cause-and-effect, to find patterns where none actually exists. In fact, when it comes to finding patterns, our species leaves the "autoshaping" of the pigeon far behind. Indeed, the intellect that permits us remarkable logical and creative powers also makes us prone to elaborate all sorts of specious associations into unholy self-deception. Like the pigeon, humans tend to irrepressibly connect aspects of their behavior with associated features or occurrences in the environment. However, unlike the pigeon, humans will also offer up a rationale for these connections. They may even conceptualize such reasoning into abstract rules that can be applied in similar situations. Finally, they may even try to convince others of the validity of these rules. This tendency appears to be, at least in part, related to the evolution of lateralization in the human central nervous system—the development of specialized cognitive functions in each of the two hemispheres of the brain. It is within the left hemisphere, neuroscientists have postulated, that the uniquely muscular human coherence-building apparatus may chiefly reside.

Beginning in the 1960s, Michael Gazzaniga was among a group of scientists who had the opportunity to study so-called split-brain patients—patients whose intractable seizure disorders necessitated severing the bulk of the white matter fibers that connect the two hemispheres. As in many neurological syndromes, the disruption of normal function and anatomy permits one to glimpse the seams of the gestalt. It is often only when regions are damaged and normal connectivity is disrupted or distorted that researchers can appreciate specific subroutines that under normal circumstances remain submerged by the well-orchestrated hum of an intact machine. This is exactly how Gazzaniga and colleagues discovered that there appears to be an "interpreter" module in the left hemisphere that creates coherence—knitting cause-and-effect, whether or not there is reliable data available. For example, when written commands were flashed to the right hemisphere of split-brain patients (unbeknownst to the isolated left hemisphere), the subject would report a *post hoc* confabulated explanation for his or her behavior. The command "walk," for example, would cause the subject to get up and move around. However, when asked about this, the subject replied that she was thirsty and had simply gotten up to get a drink (Hirstein, 2006, 154). In another instance, a subject's right hemisphere was shown disturbing, violent images. When the subject was asked what she had seen she replied that she was not sure, but complained of feeling vaguely frightened, which she then attributed to the behavior of the investigator. Gazzaniga realized that although a detailed percept of the disturbing scene could not be relayed to the disconnected left hemisphere, the negative emotional valence attached to the scene, relayed through sub-cortical intact non-collosal relays, led the left hemisphere to confabulate an explanation for the patient's emotional reaction based upon unrelated features in the environment (Hirstein, 2006, 154). In contrast the right hemisphere, although more linguistically and cognitively limited than the left, shows less of this confabulatory tendency but rather may function to maintain accurate, veridical accounts and responses to the data it has available with less elaborative distortion and coherence-building functionality. Certain right-sided regions therefore may also have important emotional inhibitory roles in the normal integrated brain and function as both epistemic and behavioral checks (Hirstein, 2006, 170).

In the intact brain, these left and right-sided capacities obviously work together (along with other self and environmental monitoring systems that are not necessarily neatly lateralized)—striking a balance between the tendency to find patterns and explanations and, on the other hand, generating a more unadorned, accurate representation of the environment and our own behavior. But the confabulatory tendency—dramatically released under pathological split-brain conditions as well as in other confabulatory neurological syndromes—still likely exerts its influence in all sorts of intriguing ways in the normal, intact brain. We can perhaps see the distorting influence of the coherence-building function of consciousness in a witness's tendency to be influenced by an overzealous investigator or in the ways certain personality types marked by impaired insight can misinterpret the intentions of others and rationalize all sorts of self-serving behaviors. Of course phenomena released by gross lesions of the brain may or may not be relevant to behavior displayed by the intact brain; furthermore, other non-lateralized forms of "causality-reading" may exist. However, this feature of information processing, unmasked under pathological conditions, likely supports the supposition that there are cause-and-effect, interpreter modules within the brain and that there may be some relationship between pathological confabulation and the ubiquitous coherence-building functions in the intact brain that are associated with all sorts of erroneous or distorted outputs under more pedestrian conditions.

So, what relation does all of this have to the sorts of narrative explanations of behavior (abetted by the Freudian revolution) that continue to this day? Obviously humans have always been engaged in coherence-building interpretations of events and behaviors. However, the Freudian revolution elaborated the causality-reading tendency into a therapeutic worldview that fundamentally shifted the way behavior was understood for generations. This is not to argue against the usefulness of introspection in general—keeping tabs on the near-term motivational vectors that influence one's choices is crucial for sound judgment. However, introspection is often of limited value in determining *ultimate* causes of enduring behavioral proclivities and pathologies.

As previously stated, Freudianism is long dead and buried in academic psychiatry. Many might therefore question the point of even discussing Freud's ideas given how obviously outdated and unscientific they are. However, subterranean cultural and clinical influences from the Freudian revolution clearly persist. As noted above, one can still see the influence of the biographical or narrative view of identity in various artistic productions, biographies, and public commentaries. Furthermore, there are still psychoanalytic institutes in most major cities and neo-analytic ideas are taught in many programs that train therapists. It is not within the scope of this book to discuss Freud's massive project in any detail; those seeking some of the better critiques of Freud could start with the work of Adolf Grunbaum, Frank Cioffi, Fred Crews, and Frank Sulloway, to name some of his most able critics. However, a few examples of Freud's explanations of psychopathology will demonstrate that the founder of the therapeutic narrative, as well as his legatees to this day, construct theories of mental pathology that devolve chiefly from a form of discursive, cause-and-effect logic that has little to do with the highly specified ways that emotions are regulated in our species. And those who would argue that it's absurd to discuss such outmoded theories need only peruse a contemporary analytic journal or a syllabus from a program that trains psychotherapists. Freud's lexicon and thematic focus are largely gone, however many of the errors endemic to biographical explanations of behavior remain strikingly unchanged.

One, then, might as well begin at the beginning. Consider, as an example, Freud's theory of the genesis of paranoia. This is a symptom that, even in the late nineteenth century, suggested to many physicians the presence of a disease process. Although Freud frequently warned about the limits of his method when dealing with psychosis, he nonetheless could not resist psychological speculation concerning the etiology of psychotic phenomena. This is not surprising as Freud tended to view all symptoms—whether mild or severe—of a kind: they all resulted from pathogenic experiences and developmental arrests. In cases of paranoia, the patient had become stranded in the auto-erotic phase of development when the self (i.e., the same sex) is taken as the loved object. Anxiety over this same-sex attraction leads to the repression of conflictual desire. Finally, a secondary defensive reaction to these unacceptable feelings leads the patient to feel persecuted by (instead of drawn to) the loved object.

Now if we switch around a few circles and arrows in the Freudian schematic, we find that another form of forbidden sexuality—incestuous love—can explain a case of lesbianism! In this instance, a young woman's love for her father is dashed by her mother's pregnancy; having been bested by her rival she turns away from all men and thus forsakes heterosexual love altogether. In another interpretative sleight-of-hand from the famous Dora case, a girl's shortness of breath is linked to having overheard her parents having sex: the experience of hearing a father's heavy breathing, in Freud's fable, forms the nidus of a stubborn case of hysterical asthma. In these fairly typical examples of the Freudian schematic, internal coherence serves as evidence of an explanation's cogency. Many of the errors that biologist Michael Ghiselin has warned are characteristic of misguided teleological thinking are exhibited in such theories: natural phenomena—in this case mental symptoms—are always assumed to display instrumentality, to serve some sort of purpose. This often leads to the conflation of utility (i.e., the purported psychological function of a symptom) with the ultimate cause or origin of the symptom or disorder in question. Additionally, such explanations often follow the contours of an abstract, disembodied logic (Ghiselin, 1974).

As a number of Freud's critics have documented, many of these etiological theories are shockingly contrived—see the work of Frank Cioffi for some of the wittiest depictions of the florid promiscuity of Freudian hermeneutics (Cioffi, 1998). The wide variety of mental pathologies are attributed not to the variegated aspects of emotional regulation; rather Freud puts forward a universal explanation for mental infirmity based upon conflictual desire which becomes deformed and partially submerged by the mind's defensive operations. Variation in specific symptoms is supplied by the particularities of historical events, memories, and the particular mental defenses employed. According to this confabulatory-historical method, mental symptoms need only to be traced back to signal events to reveal the secret of their inception. Now in order to result in mental illness, these historical events had to be fairly powerful; therefore the sexual instincts were recruited to provide sufficient motive force.

Perhaps now the reader can appreciate how Freud's thinking can resemble the "superstitious" behavior seen in animals as he reads causation from proximity. But in Freud's case, this proximity is thematic rather than temporal or spatial: pathological emotion is inevitably linked to thematically related pathogenic experience. But Freud's thinking (and the placing of the reproductive instincts at center stage) also reflects another innate human tendency—our species' preoccupation with impurity. In the Freudian narrative, as in many religious traditions, infirmity and suffering are tied to what are deemed impure acts or transgressions of natural law—in Freud's case these transgressions involved incestuous desire and non-reproductive sex (masturbation and homosexuality). Ironically, the man who supposedly liberated the West from Victorian prudery by placing sex at the center of human psychology, promulgated theories that were fundamentally moralistic and regressive. As Gore Vidal once opined, *Leviticus* and Freud's *Standard Edition* turn out to be bookend texts.

It is ironic, then, that such theories were taken up by the French avant-garde who, in the 1960s and 1970s, attempted to launch a

Freudian renaissance in the French academy. A number of French psychoanalysts, chief among them Jacques Lacan, became intellectual superstars. Unfortunately Lacan put forward theories of psychopathology that lacked any credible link to science or clinical psychiatry. Instead, he fetishized the patient's discourse as a linguistic rebus—the ultimate narrative puzzle—requiring the most recondite mathematical and quasi-mystical decoding. In one fairly typical inscrutable passage, Lacan likens the male reproductive organ to the square root of negative one. Unfortunately, Lacan was no passing fad. To this day there are corners of academia that remain devoted to his antics. In fact, two accomplished physicists were so dismayed by Lacan's ongoing influence that, in the late 1990s, they included a section on his work in their book, *Fashionable Nonsense: Postmodern Intellectuals' Abuse of Science*. This scathing compendium of fraudulence and obscurantism chronicles psychoanalytic flirtations with mathematics and physics. The authors concluded that Lacan's misappropriations of various mathematical and scientific ideas were so incoherent as to be largely incomprehensible to those with formal training in these disciplines (Sokal and Brichmont, 1998).

But leaving aside the idiosyncrasies of French intellectual fashion, the contemporary legatees of the talking cure have, of course, long ago jettisoned Freud's baroque, outlandish sexology. In its place, neo-psychoanalytic theories emerged. One of the most prominent among these is known as "attachment theory." The story has been transposed into another key but it is still storytelling nonetheless: we have become who we are because mental structures are derived from a noetic process whereby experiences and ways of relating are internalized. In attachment theory there are still historical pathogenic insults and developmental logjams, but these have nothing to do with sex; rather they are now tied to inadequate early social bond formation with caregivers. According to these theories, the capacity to "mentalize" (to differentiate and understand the feelings of oneself and others) is not innate but must be developed in early childhood through proper interaction with parents. This is accomplished through a process described as "affect mirroring" by which the parent reflects and relays affect states to the child. As this mentalizing capacity is deemed critical in learning how to regulate one's emotions, poor affect mirroring in childhood is associated with poorly regulated affects and problems with attachments and relationships later in life (Pedersen et al., 2015).

A theory such as this—that ascribes formative primacy to the parent–child bond—has predictably been recruited to explain an expanding range of disorders including eating disorders, various personality disorders, and even gender identity disorder. Although most writings on attachment theory embrace twenty-first-century nomenclature and, on the surface, sound less outlandish than offerings from prior generations, the theoretical framework of much of this material maps with great congruence upon the original Freudian schematic; furthermore, such work often bears little relation to other fields of knowledge except those of an equally hermetic nature. To be fair, there may be aspects of attachment theory that might be relevant to some problems—for example the difficulties found in children raised in orphanages under conditions of severe social deprivation. However, there is much to be skeptical about. In recent decades, when science finally has worthwhile things to say about emotion and has established significant authority with the public, some psychoanalysts perhaps needed to beat a path back to biology and the brain. They have found this path, apparently, by focusing (just as Freud had done) on a set of instinctual behaviors. But this time, the focus is not on the reproductive instincts but on what Darwin referred to as the social instincts, such as those involved in social bond formation: *voilà*, sex has been replaced by attachment. This trend may follow from analysts' special interest in those patients with particularly difficult interpersonal lives—that is, patients with personality disorders whom modern psychiatry cannot easily help with medications and who often experience distress related to impaired interpersonal function. However, as will be discussed in Chapter 8, some studies have indicated that many of the personality disorders are almost as heritable as mood and anxiety disorders. But because the traits that underlie these disturbances appear to be too diffuse and complex to be tied to modular neurobiological systems (and are often difficult to treat), the analytically oriented clinicians and theorists have traditionally taken center stage in this arena. Now to be sure, there are plenty of skilled contemporary therapists who treat personality disorders with commonsense, empirically tested approaches that help patients understand their destructive proclivities and assist them in trying to tolerate the affects and anxieties that propel maladaptive behavior. As reiterated above, psychotherapy can be very helpful for many patients. However, a therapist's skill is likely most often chiefly a function of the therapist's own innate emotional intelligence, their understanding of critical dimensional traits, and, of course, their level of clinical experience and professionalism. Complex, didactic structures or generalizable theories concerning the origin of various types of pathology often add little if anything to a therapist's toolbox; indeed, the practice is far more craft than science.

Finally, attachment theory notions of mentalizing/reflective functioning are reminiscent of the theories of a group of autism researchers who have argued that people with autism appear different from a very early age because they *congenitally* have deficits in their ability to acquire an adequate "theory of mind"—the capacity to perceive the intentions, beliefs, or emotions of others (Baron-Cohen et al., 1985). Autism is in fact an excellent example of how certain disorders, by impairing a function, actually demonstrate that the function in question—a function that attachment theorists claim is developed through interaction with the parent—may in fact be more a product of innate endowment than environmental induction. Indeed when one looks at the most basic social-emotional tasks that many with autistic spectrum disorders cannot perform, it becomes patently obvious that these functions represent innate dimensions of behavior. That is, they are not amenable to psychological explanation but rather are linked to core biological endowments. Thus, for children growing up in families not characterized by significant deprivation or abuse, processes resembling "mentalizing," likely develop as part of the child's natural repertoire of cognitive/emotional capacities. However, many therapists, like the superstitious pigeon, are still prone to the reflexive reading of cause and effect: persons with problems regulating their affects and managing attachments in adulthood must have had caregivers whose inadequacies impaired the development of normal attachment and emotional regulation during childhood. Hence, offshoots and permutations of analytic theory persist to this day. Despite the changing theoretical and thematic focus, a central supposition remains: one acquires and internalizes mental structure and maladaption from one's early environment and caregivers.

References

Baron-Cohen, S., Leslie, A.M., Frith, U. (1985). Does the autistic child have a theory of mind. *Cognition*, 21, 37–46.

Cioffi, F. (1998). Claims without commitments. In Crews, F. (Ed), *Unauthorized Freud: doubters confront a legend*. New York: Penguin Group.

Ghiselin, M. T. (1974). *The economy of nature and the evolution of sex*. Berkeley: University of California Press.

Hirstein, W. (2006). *Brain fiction: self-deception and the riddle of confabulation*. Cambridge: MIT Press.

Panksepp, J. (1998). *Affective Neuroscience*. New York: Oxford University Press.

Pedersen, S., Poulsen, S., Lynn, S. (2015). Eating disorders and mentalization: high reflective functioning in patients with bulimia nervosa. *Journal of the American Psychoanalytic Association* 63, 4, 671–694.

Sokal, A., Brichmont, J. (1998). *Fashionable nonsense: postmodern intellectuals' abuse of science*. New York: Picador.

Part II

Apostles of Modern Nativism

This part contains discussions of important figures from science, medicine, and literature whose insights into innate influences on behavior were particularly prescient and ran decidedly contrary to the modern declivity toward the self as narrative.

3 Francis Galton and the Birth of Behavioral Genetics

I have often wondered when the idea first took hold of him: that the lineaments of our destinies lay buried in our forbearers' graves; that one is born, in a sense, fully formed according to the laws of a yet undiscovered science—the particulars of which would soon be illuminated by the Moravian beekeeper Gregor Mendel. Perhaps it was during that dreary November in 1840 while he was a student at Cambridge. Francis Galton was then a callow first-year but he knew one thing for certain: he was quite unwell. The rigors of academic life (mathematics in particular) had caused him to take to bed—the first in a series of mental collapses. Perhaps it was then that he first began to wrestle with the enigma of temperament and its transmission through the generations. Maybe it was the shape of his thin, tight-set mouth that reminded him of a close relation—one who had also suffered bouts of nervous exhaustion. Perhaps immobilized in a bath growing lukewarm he thought: someone who came before me must have experienced this exact cast of mind. He would recover and go on to complete his degree but the riddle of inheritance would occupy him for a lifetime. The youth, who was driven to bed by the calculus, would go on to become the founder of the discipline of biostatistics, a close collaborator of Darwin, and a prodigiously inventive scientist in his own right.

Today, however, Francis Galton is virtually unknown outside of scientific and historical circles. Among his many contributions were the forensic use of fingerprints, the concept of regression to the mean, and the first empirical studies of twins. But Galton's greatest achievement remains his prescient ideas on the inheritance of behavioral traits. Long before there were credible theories of the mechanism of heredity, Galton set out to prove that mental traits, talents, and pathologies were as heritable as curly hair and a high forehead. The modern discipline of behavioral genetics—which uses molecular and population

DOI: 10.4324/b23263-6

approaches to parse the puzzle of human temperament—emerged from the work of this shy, eccentric Victorian.

Galton's relative obscurity may stem from his role as a pioneer of eugenics—the effort to ameliorate social and medical ills by influencing who gets to reproduce. Given the moral hazards inherent to eugenics (as well as its more sinister fellow-travelers later in the twentieth century), it is not surprising that Galton's stature suffered as a result of his political views. To make matters worse, his intellectual gifts were alloyed with decidedly odd personality traits—including an obsessive drive to quantify the most trivial variables. Indeed, his was a career that encompassed extremes of inspiration as well as fatuity: he developed methods that would form the foundations of population genetics, but he also undertook experiments on tea brewing, gave hearing tests to the occupants of London Zoo, and was the author of an article entitled "Three Generations of Lunatic Cats" (Galton, 1896). In his personal life he was, as biographer Martin Brookes described, "an immense snob" who wished to turn marriage into a "rutting club" for the elite (Brookes, 2004, 47).

Belief in selective breeding was not, strictly speaking, a nineteenth-century notion. One can find evidence of it embedded in cultural practices throughout the world and across the centuries. The conviction that children will carry forward the pathologies in their families of origin may underlie, along with economic considerations, the custom of arranged marriage—a custom that may constitute the world's oldest and most common eugenic practice. The encyclopedic Renaissance treatise on mental infirmity, Robert Burton's *The Anatomy of Melancholy* (1621), illustrated the prevailing, pre-scientific assumption that temperament and mental illness were inherited; it also contained arguments in favor of selective breeding:

> For now ... in giving way to all to marry that will ... there is a vast confusion of hereditary diseases .... It comes to pass that ... we have many feral diseases raging amongst us, crazed families, *parentes peremptores* [our parents are our ruin], and our fathers bad, and we are likely to be worse.
>
> (Burton, 2001, 211)

Galton, then, was following in an age-old tradition of viewing mental traits no differently from physical ones. Perhaps Galton should not be too harshly judged for his faith in eugenics. The Victorians had no effective treatments for mental illness. Tallying up the incidences of the major mental disorders quickly reveals the tremendous burden that these maladies placed on families and communities. Because of this, enthusiasm for eugenics was not restricted to conservatives. George Bernard Shaw was a fan of Galton's work and viewed eugenics as essential for the preservation of civilization. In this light, Shaw's literary send-ups of human courtship can be viewed as cautionary tales revealing romantic love to be a flawed approach to reproduction. Infatuation—that paragon of human folly—surely offered the playwright a broad comedic target on many levels; but, from a trans-generational perspective, Shaw viewed it as something akin to entrapment. Like Galton, Shaw looked forward to a more rational system of human reproduction. The fictional narrator's tongue-in-cheek comments in Shaw's *The Revolutionist's Handbook and Pocket Companion* were probably not far removed from his own views: "Even a joint stock human stud farm (piously disguised as a reformed Foundling Hospital or something of that sort) might well, under proper inspection and regulation, produce better results than our present reliance on promiscuous marriage" (Shaw, 1903, 254).

It is more accurate, then, to regard Galton guilty not of political extremism but scientific idealism. As evinced by Shaw's comments, this is a charge that could have been leveled at those on the left as well as the right in the late nineteenth century. Furthermore the more egregious and truly horrific policies connected with the eugenics movement occurred after Galton's death. As one of his biographers, Nicholas Gillham, has argued, Galton was not a callous or cruel man and he would likely have been shocked and dismayed at many of the policies—such as forced sterilization—that resulted from the efforts of some of his followers (Gillham, 2001, 357). But apart from his role in the eugenics movement, Galton also played a significant role in the Darwinian revolution by developing statistical approaches to inheritance and by undertaking the first empirical studies of the inheritance of mental traits.

Galton realized that the inheritance of complex traits appeared "capricious" at the level of individual families. Therefore clues to the mechanics of heredity could only be revealed through a change in scale—by scrutinizing patterns of inheritance in populations. Galton subjected all sorts of traits to statistical analysis: height, arm strength, the speed at which word associations were generated. But his main interest, as a eugenicist, was charting variation in human mental traits and cognitive abilities. In his paper "Hereditary Talent and Character" (1865), Galton tried to prove that intellectual gifts were inherited by demonstrating that talent clustered in families. By studying persons of exceptional talent, in fields least impacted by nepotism, Galton created a statistical lever that could yield significance despite a thicket of confounding variables. Like a lot of Galton's work, this paper contained methods that were crude yet inventive and the results are difficult to attribute to chance alone. He estimated that Europe had produced 330 individuals of prodigious literary and scientific genius over the preceding four centuries out of a total of approximately one million highly educated persons. This yielded a "genius incidence" of roughly one in 3000. However, of these 330 geniuses, 51 (approximately one-sixth of them) also had a close blood relative with exceptional gifts. This represented an odds' ratio of 465—that is, a close relation of a creative or scientific genius would be 465 times more likely to possess great talent than any unrelated but highly educated individual (Galton, 1865).

Although Galton focused on population-based methods to prove the heritability of mental traits, he understood that any workable theory of heredity also had to be able to account for individual exceptions to the trend seen in groups. Why, for example, did two gifted parents frequently produce unexceptional offspring? In fact, as Galton observed, parents who showed extreme variation in a particular trait often produced children who showed less variation from the mean than their parents. He called this statistical phenomenon "regression to the mean" and postulated that it arose because of the way heredity operated—namely, children were not a blend of their parents' features but instead derived their traits from a population of ancestors stretching back over the generations. The collective genetic legacy of these ancestors pushed the offspring of outliers back toward the mean.

Galton's effort to explain regression to the mean led him to a related discovery—his theory of the ancestral law which posited that offspring often possessed traits not found in their parents but which were present in more distant ancestors. This implied a distinction between phenotype and genotype—a distinction that the Danish botanist Johansson more explicitly elucidated. Galton's concept of ancestral inheritance was prescient in other respects as it implied that the germ cells contained a storehouse of particulate information which made its way down the generations unaltered—sometimes going unexpressed, other times popping up in an unpredictable fashion. Accordingly, offspring were not so much the product of two parents but rather the descendants of a chain of ancestral embryos (Galton, 1889).

Galton's ideas on inheritance pre-dated the rediscovery of the work of Mendel, yet his speculations were quite insightful for a man stumbling about in the dark. His notion of units of heredity—purveyors of information that descended essentially from embryo to embryo, that could remain latent or be expressed, and that were segregated from somatic influence precisely because they were fundamentally of a different nature than the structures they instantiated—prefigured aspects of August Weissman's theory of the "continuity of the germ plasm," which is generally regarded as the first purely informational concept of inheritance (Bulmer, 2003, 133).

Galton's work was also influenced by Darwin's work on animal behavior. Both men viewed the species-specific behaviors found in various animals as proof of the heritability of highly complex behavioral repertoires. Galton was particularly interested in the social behavior of domesticated animals and cited the limited number of species that could be domesticated as proof of the innate nature of behavior:

> only a few species of animals are fitted by their nature to become domestic, and ... these were discovered long ago .... All the suitable material whence domestic animals could be derived has long since been worked out .... The finality of the process of domestication must be accepted as one of the most striking instances of the inflexibility of natural disposition.
>
> (Galton, 1907, 174)

In other words, if a species lacked the rudiments of certain prosocial behavioral traits crucial for domestication (i.e., a degree of docility and/or cooperative, affiliative tendencies) then all the coaxing and training would come to naught. As will be discussed in Chapter 6, domestication only works because a change in conditions (living with humans) reveals stores of previously latent genes that can be exploited through animal husbandry. If certain genes underlying critical behavioral proclivities are completely lacking in a species (like the zebra, for example), then there is no basis for selective culling and breeding.

To Galton, people were fundamentally no different from oxen: some were pleasant and agreeable; others were obstinate and quarrelsome. Long before the scientific age, people had no trouble distinguishing the various dimensions of temperament and the basic human personality types. Although these types were easy enough to spot, the empiric study of personality and its inter-generational transmission was certainly beyond crude nineteenth-century science; true to form, Galton dove right in. In his 1887 study, "Good and Bad Temper in English Families," Galton gathered descriptions of informants' family members, which he had obtained by survey, and turned these descriptions into categorical data in order to measure the correlation of temperament within families.

Many of the negative traits approximate the modern psychiatric criteria for the more severe personality disorders with an added sprinkling of mood symptoms. The ill-tempered were described as "acrimonious, rageful, arbitrary, capricious, impetuous, easily-offended, despotic, irritable, vindictive" (Galton, 1887, 22). In short, this group represented that 5% of the population that, while not acutely mentally ill, makes the rest of humanity miserable. Galton believed that these sullen, ungovernable people were the scourge of domestic tranquility—adding to the societal ills of violence, pauperism, and debauchery.

He predicted that looking across many families would reveal patterns of inheritance that were not necessarily apparent when examining individual families due to the tendency of offspring to display traits inherited from more distant relations. Of the approximately 70 families he studied, he found that good-tempered offspring were three times more likely to emerge from good-tempered parents than from bad-tempered parents. From the union of two bad-tempered parents, there was a greater than ten to one ratio of bad to good-tempered children.

Galton knew that studies of temperament were open to criticism given the difficulties of controlling for the influence of environment. He realized that twins represented a unique opportunity in this regard. In 1875 he undertook the first twin study described in the paper "The History of Twins, as a Criterion of the Relative Powers of Nature and Nurture." Despite a purely anecdotal approach, a lack of access to twins raised apart, and Galton's poor understanding of the distinction between mono and dizygotic twins, the paper is filled with uncanny insights. Galton followed the development of 35 pairs of identical twins and 20 pairs of dissimilar twins who, like regular siblings, had quite distinct temperaments. He found that the identical twins remained extremely similar in temperament even as they grew older and were exposed to varying influences outside the home. As borne out in modern twin studies, he discovered that identical twins actually seemed to converge as they aged as time brought out their temperamental vulnerabilities (Galton, 1875). Regarding the impact of a shared environment upon the dissimilar twins, he found that without exception there seemed to be little convergence in those born with distinct temperaments despite being subject to the same school and home environments. He stated

> there is no escape from the conclusion that nature prevails enormously over nurture when the differences of nurture do not exceed what is commonly to be found among persons of the same rank of society and in the same country.
>
> (Galton, 1875, 404)

Interestingly, in the lines above Galton appeared to give a passing nod to the impact of environment by noting that it will count for little *as long as* it is reasonably typical—therefore implying that exceptional environments might indeed have a significant impact. Remarkably, Galton also prefigured contemporary ideas about the dynamic interaction between genetic and environmental influence:

> Much stress is laid on the persistence of moral impressions made in childhood, and the conclusion is drawn, that the effects of early teaching generally, must be important in a corresponding degree. I acknowledge the fact, but doubt the deduction. Its parents usually teach the child, and their teachings are of an exceptional character, for the following reason. There is commonly a strong resemblance, owing to inheritance, between the dispositions of the child and its parents. They are able to understand the ways of one another more intimately than is possible to persons not of the same blood, and a child instinctively assimilates the habits and ways of thought of its parents. Its disposition is educated by them, in the true sense of the word; that is to say, it is evoked earlier than it would otherwise have been.
>
> (Galton, 1875, 405)

Here Galton appears to anticipate the contemporary notion that one's genetic endowment plays a role in one's experience of the environment; in fact, genetic endowment may actually influence one's environment to a significant degree—a process behavioral geneticist Robert Plomin has dubbed "the nature of nurture." For this reason, Galton intuited that close proximity with blood relations during development may elicit innate proclivities at an earlier age.

Remarkably, Galton's intuitions in these matters have been empirically borne out. Many measures of environmental factors actually show significant levels of heritability. Surprisingly, even discrete stressful life events (traditionally considered to be among the most paradigmatic of environmental inputs) appear to owe almost a third of their variance to genetic antecedents (Plomin, 2018). In other words, one's genetic endowment plays a significant role in eliciting or influencing even fairly discrete extrinsic phenomena: think of the increased number of accidents that often befall more impulsive persons, or how children with certain social impairments are frequently bullied at higher rates. In these examples, the genome is exerting an indirect but significant effect on the environment. These findings might also explain why monozygotic twins tend to show greater similarity as they age and dizygotic twins less so: it would seem that the genetic endowment has had more time to incubate its own good and bad fortune—leading to more convergence with the accumulation of apparent happenstance, happenstance that apparently carries the indirect influence of one's temperament.

Darwin himself eventually weighed in on the relative contributions of innate and extrinsic influences, indicating in his autobiography that he agreed with Galton that psychological traits were largely innate, environment playing only a minor role (Bulmer, 2003, 57). Fortunately Darwin's stature and legacy were not impacted by his eccentric cousin's more extreme hereditarian enthusiasms—tainted as they were by elitism and racism. Nonetheless Galton's remarkable insight and ingenuity lent scientific credence to what common people have intuited for millennia—namely, that human offspring closely resemble their forbearers behaviorally as well as physically. Although Galton's era witnessed both the apogee of the novel as well as the birth of "the talking cure," this quirky Victorian decided that, in certain domains, narrative was decidedly overrated.

References

Brookes, M. (2004). *Extreme measures*. New York: Bloomsbury Publishing.

Bulmer, M. (2003). *Francis Galton: pioneer of heredity and biometry*. Baltimore: Johns Hopkins University Press.

Burton, R. (2001). *The anatomy of melancholy*. New York: The New York Review of Books.

Galton, F. (1865). Hereditary talent and character. *Macmillan's Magazine*, 12, 157–166.

Galton, F. (1875). The history of twins as a criterion of the relative power of nature and nurture. *Journal of the Anthropological Institute*, 5, 391–406.

Galton, F. (1887). Good and bad temper in English families. *Fortnightly Review*, 42, 21–30.

Galton, F. (1889). *Natural inheritance*. London: Macmillan.

Galton, F. (1896). Three generations of lunatic cats. *Spectator*, April, 514–515.

Galton, F. (1907). *Inquiries into human faculty and its development*. London: J.M. Dent and Sons.

Gillham, N. W. (2001). *A life of Sir Francis Galton: from African exploration to the birth of eugenics*. Oxford: Oxford University Press.

Plomin, R. (2018). *Blueprint: how DNA makes us who we are*. Cambridge: The MIT Press.

Shaw, G. B. (1903). *Man and Superman: a comedy and a philosophy*. Baltimore: Pequot Books.

4 The Novelist as Accidental Nativist
The Addict as Natural Kind

This chapter takes a detour away from science and consulting rooms into the contrivances of nineteenth-century fiction. The aim is to enlist the narrative arts in this polemic against narrative. It will be argued that although Honoré de Balzac did not explicitly traffic in nativist ideas, he can be considered a crypto-nativist for a number of reasons. First off, his characters display a remarkable degree of behavioral stasis: the patterns of choice in Balzac's fictional world appear fixed and immutable. As a devotee of Spinoza, he maintained a fairly deterministic view of behavior and was likely skeptical of free will. Consistent with this, a number of his characters explicitly opine upon the inborn and inflexible nature of behavior. In Balzac's fiction, as in Greek drama, identity is not forged by destiny, rather it *is* destiny. Finally, Balzac showed a particular preoccupation with behaviors and traits that he clearly believed ran in his own pedigree. Balzac's implicit nativism supports the contention that our contemporary therapeutic culture—which still cleaves to biographical accounts of behavior—is in fact an historical aberration.

Balzac is considered one of the founders of literary realism. His massive *La Comedie Humaine*, composed of nearly 100 novels, attempted nothing less than a comprehensive rendering of post-Napoleonic French society. But perhaps the most important achievement of the early modern novel was a new level of behavioral realism—and Balzac achieved an uncanny level of psychological verisimilitude. In fact, many of his depictions of the behavioral traits of his characters are so congruent with contemporary categories that they could be said to provide a degree of historical construct validity. Such depictions add credence to the idea that particular constellations of traits represent natural kinds—recognizable even through the lens of time and culture. Balzac, as mentioned above, was also especially interested in traits and pathologies that ran in his own family. In fact, one of the

DOI: 10.4324/b23263-7

main characters in his novel *Cousin Bette* was likely modeled after Balzac's father—the man from whom he had inherited an impulsive, addiction-prone temperament.

Balzac clearly understood that the addict does not necessarily drink to excess, use laudanum, or keep a snuffbox filled with hashish. He knew, from personal experience, that the addict is not simply one who falls under the sway of a dangerous substance: the addict is *born* different—ruled not so much by the allure of intoxicants but by his own impulsive nature. It is not surprising that he modeled the villain of *Cousin Bette* after the singularly distasteful father whose genetic legacy wrecked havoc in Balzac's personal and professional life. The realism he achieved, therefore, may owe as much to his insight into the inherited underpinnings of his own behavior as to his appreciation of the idiosyncrasies of French society. The core dimensions of addictiveness that he so lucidly described include: high levels of impulsivity (the prioritizing of immediate over delayed rewards and the tendency to discount risk); the prominent deployment of suppressing or detaching mental defenses; and, finally, a susceptibility to the loss of behavioral control in response to repeated exposure to certain categories of reward.

A striking loss of self-control in the face of the most dire consequences is what makes addiction such a vexing and dangerous disorder. But this uniquely destructive human tendency has illuminated important truths about the mechanisms that control motivated behavior in all animals. Neuroscientist Jaak Panksepp has aptly named this collection of functionally related mechanisms the "seeking system"— the ancient neural circuitry responsible for reward-related learning and pursuit. Balzac clearly understood that there were those of a particular mental cast who were easily enslaved to rewards (like sex) just as the drug addict becomes captive to intoxicants. And he clearly distinguished this pathological type from the garden-variety womanizers, rounders, and spendthrifts who flit in and out of *The Human Comedy*. There is little doubt that Balzac was intrigued by the so-called behavioral addictions, such as those involving gambling or compulsive sexual behavior. He also clearly viewed these disorders as variants of classical addiction.

The precise relation of behavioral disorders involving so-called natural rewards such as sex and food to traditional substance addiction remains an area of research and debate; however, no serious student of addiction doubts the significant overlap of these phenomena. Both of these categories show clear substitutability: for example, a significant number of patients who undergo bariatric surgery for compulsive eating develop what is called addiction-transfer—acquiring new substance as well as non-substance addictions once their overeating is controlled (Blum et al., 2011). It is also well known that Parkinson's disease medications—that sensitize the same reward-related processes that are primed during addiction—regularly induce *de novo* behavioral addictions such as gambling and compulsive sexual behavior (Olsen, 2011). Animal studies also suggest that the protracted neuroadaptive changes that occur in the brains of rats after repeated exposure to natural rewards closely resemble those changes induced by repeated exposure to drug rewards; additionally, such studies have also demonstrated bidirectional cross-sensitization between natural rewards (access to sex or sucrose) and drugs such as amphetamine (Olsen, 2011).

The natural rewards are obviously the reinforcers that the seeking system evolved to motivate animals to pursue. Intoxicants, then, are merely serendipitous tag-alongs that exploit the natural reward circuitry of the brain. Therefore the traits that underlie addictive behavior existed in populations long before agriculture permitted the discovery of fermentation, and long before the invention of money—that most fungible of devices that can provide ready access to sex, goods of all sorts, and gaming in all its varieties. If capital can be considered the ultimate foraging tool, then acquisitiveness is the mode of seeking that most typifies the post-industrial age. Balzac was arguably the greatest chronicler of human acquisitiveness because he himself suffered from a type of compulsive collecting that was more akin to gambling than consumption.

Accordingly, the so-called "cousin novels" (*Cousin Bette* and *Cousin Pons*) are rife with references to gambling, billiards, compulsive collecting—of everything from furniture to artwork to courtesans. But one should not be fooled, as Marx was, that Balzac's depiction of vice and excess indicated a concern with the corrosive effects of industrialization and mercantilism. In fact, such themes in Balzac's work had nothing to do with enlightened political sympathies—which Marx and Engels had imagined Balzac shared—and everything to do with Balzac's personal financial struggles. Balzac was actually politically conservative; however, his compulsive spending and chronic debt created an ambivalence toward wealth that apparently confused Marx. This ambivalence is mirrored in his treatment of the figures that inhabit *The Human Comedy*: he lampoons their material ambitions while lavishing fetishistic attention on their possessions. If capital is the invisible third character in much of mid- to late-nineteenth-century fiction, then for Balzac it must be considered the central figure—whether disguised as a harlot on-the-make or a nouveau riche merchant.

Cousin Bette is a shockingly obscene book for its time. The excesses of the central figure, Baron Hulot d'Ervy (the villain who was modeled after Balzac's father), are jaw-dropping even for the modern reader accustomed to graphic depictions of all manner of coupling. It is difficult to imagine why Flaubert was charged with indecency for writing *Madame Bovary*, while Balzac got off scot-free. A brief summary of Baron Hulot's adventures will help orient the reader. The Baron is an aging aristocrat with a position at the War Ministry. At the onset he appears to be a typical libertine. Like many of his class, he has enjoyed liaisons with a number of women including a singer, Josepha, whom he has stolen from Crevel—an ex-perfumer and a striver (when Balzac hates a character, and he appears to despise Crevel, he reliably calls him a "shop-keeper"). Now this shop-keeper has managed to purchase respectability and a few civic titles but he could not out-bid the Baron for possession of Josepha, "slight and wiry, with the golden skin of an Andalusian ... the style of a duchess ... and the pretty ways of a wild faun." When the Baron steals Josepha, Crevel becomes enraged. He complains that, thanks to Hulot, his treasure has "been degraded to a mantrap, a money-box for five-franc pieces" (Balzac, 1909, 10). To avenge his loss he reveals the Baron's exploits to the latter's long-suffering wife, Adeline—detailing how the Baron has impoverished her family by spending a fortune on various mistresses. Crevel then makes Adeline an outrageous offer: if she agrees to have a liaison with him, he will save her family from bankruptcy and provide a dowry for her unwed daughter. She of course refuses. The Baroness later comes to regret her decision; as we shall see, virtue in Balzac's fictional universe can be as irrational and inflexible as vice.

Soon enough the Baron loses Josepha after she completes her fleecing of him. In his despair he renounces his life of profligacy and promises his wife that he will be worthy of her loyalty. But the Baron does not simply have a wandering eye: in today's vernacular he is an inveterate sex addict. Like any addict, the Baron is devoted to abstinence only as long as the next fix is out of reach. So, after briefly nursing his wounds, the Baron stumbles across the path of another professional "outrage"—Madame Valerie Marneffe—and he is at it once again. This time he goes even deeper into debt and manages to lure an aging uncle into criminality to support ever more desperate efforts to keep Marneffe. Having bankrupted his family, caused the suicide of his uncle, he only gives up on the conniving Marneffe when she colludes in his arrest on indecency charges. With her family at the brink of ruin, Adeline, in a brilliant bookend scene to Crevel's earlier attempt at debasing her virtue, offers herself tearfully to Crevel for

200,000 francs. But Crevel cannot be aroused by an act of familial sacrifice and he rebuffs her.

With Marneffe out of reach, the Baron returns yet again, hang-dog and superficially penitent to his wife. Remarkably, the family treats him like a sick man, clearly buying into the contemporary, disease model of addiction. But they are hopeful that the septuagenarian, now sequestered away and surrounded by a loving family, will finally be safe from his disordered appetites. They are mistaken. He soon flees to seek out a series of concubines. He adopts an assumed name and sets up shop in a squalid district of Paris to maintain access to women newly imported from the provinces. Adeline eventually discovers his whereabouts while doing charity work with prostitutes. In a filthy back alley she finds the Baron, attached to yet another young girl. Trapped, the Baron once again agrees to return home but not before asking Adeline—in perhaps the most outrageous scene in nineteenth-century fiction—if he "can take the girl" (Balzac, 1909, 284).

Once separated from the prostitute and back at home, Hulot appears reconciled to a peaceful dotage. The family has been saved from ruin; even the Baron's pension has been salvaged. Just as the reader breaths a sigh of relief, the family makes a critical error of judgment: they allow a woman of child-bearing age, an Alsatian kitchen maid, to bunk within striking distance of the Baron. Like a shot off a shovel the Baron is out the back door with her, never to be seen again. Adeline, the greatest enabler in the history of world literature, promptly dies of grief.

Turning to the core characteristics of addiction that Balzac so accurately depicted, perhaps the essential feature of addiction is impulsivity. Impulsivity can be defined as the tendency to respond to near-term versus delayed reinforcers. All animals, including humans, perform what is known as temporal discounting—meaning they discount the value of a reward that is delayed: a pigeon always prefers one ounce of food immediately to four ounces delayed by just a few seconds (Rachlin, 2000). Addicts, however, discount the future more steeply than normal people—they are, in effect, less able to bind themselves to longer-term perspectives although, generally speaking, decisions based on wider perspectives result in choices of greater utility. But when delays are built into decision-making (when all choices can be held at a greater remove), animals, including humans (including even the addict), display what is known in behaviorism as "ambivalence" and their preferences begin to shift away from the impulsive choice and toward choices of greater long-term utility. More complex forms of ambivalence result when particular acts conflict with more abstract choices or choices that are extended over time. An example of this is the ambivalence faced by most alcoholics: most prefer a life of sobriety or to be able to drink with moderation. However, on any given night, they prefer to drink than to embark on the long, difficult road to normalcy. Because it is always today, and sobriety is an abstraction that exists only in extended futurity, that the alcoholic chooses to drink on any given night (Rachlin, 2000, 57).

Balzac explicitly renders the features of an addict's behavioral ambivalence in his descriptions of Hulot's shifting choices. For example, when a courtesan is temporarily unavailable to him, he is able to appreciate the destructive consequences of his behavior; he appears committed to self-control and renounces his vice. However, as soon as he has access to the object of desire, the inflection point of ambivalence shifts and he reverts to the more impulsive choice. As Hulot himself laments,

> Your uncle ... is in difficulties, and it is I who dragged him there, for he accepted bills for me to the amount of twenty thousand francs! And all for a woman who deceives me, who laughs at me behind my back, and calls me an old dyed Tom! ... [And yet] I cannot resist—I would promise here and now never to see ... [her] again, but if she wrote me two lines, I should go to her, as we marched into fire under the Emperor!
>
> (Balzac, 1909, 34)

Despite Hulot's awareness of the distress his vice causes him, despite his conscious repudiation of it, he feels powerless against its irresistible pull whenever he gains proximity to it. Just as in classical addiction, the Baron's experience, apart from the run-in phase, has little to do with pleasure but rather reflects the entrainment of an aberrant, inexorable drive that is largely independent of the intrinsic qualities or the subjective experience of the object—it represents a veritable hypostasis of reinforcement. Although in its initial stages addiction is often induced by rewards with prominent hedonic properties, a simple hedonic model cannot explain addiction; in fact, the subjective experience of pleasure is not even necessarily a feature of addiction. We know this to be the case because some rewards, such as nicotine, deliver little to no subjective pleasure yet can induce powerful and durable seeking behavior. Furthermore, seeking—as a highly aroused anticipatory state—is quite distinct from the quiescent pleasures of consummatory states. Thus, for both the classic addict and the so-called behavioral addict, rewards may have intrinsic hedonic properties (heroin, food, sex), highly unpredictable, enlivening properties (gambling), or, as in the case of nicotine, no obvious subjective correlates of its reinforcing effects.

Another proclivity characteristic of this natural kind explicitly rendered by Balzac is the excessive deployment of avoidant or detaching mental defenses. This trait is often associated with a tendency to ignore or discount risk or uncertainty when making decisions. Some have argued that this may represent a sub-type of impulsivity, perhaps dissociable from the above noted abnormalities in temporal discounting. In any case, it has been demonstrated that most impulsive people, including addicts, tend to discount riskier more unpredictable rewards less than normal people (Green and Myerson, 2004). We see this, for example, when a gambler or speculator minimizes the odds against a high-risk wager or investment. Balzac explicitly illustrates this behavioral dimension of the addict when he shows the Baron taking repeated financial and professional risks to keep Marneffe while ignoring the high likelihood that these efforts will fail spectacularly. In a particularly reckless scheme, the Baron sends a pliable, elderly uncle off to Algeria to siphon money from the sale of military supplies to please Marneffe—betraying family and country in one fell swoop. For all intents and purposes he is writing the old man's death warrant. When the unholy mess is finally revealed to Hulot's superior, Prince de Wisembourg, he is dumbstruck by the transformation of this once loyal and responsible official. The prince realizes that this is not simply a botched bagatelle with an unruly mistress, or an evening-tide infatuation, rather this is the behavior of a disordered mind completely unmoored from contingency:

> "How could you," the Prince rails at Hulot, "you who knows the precise details with which in French offices everything is written down at full length ... you who has so often complained that a hundred signatures are needed for a mere trifle .... How could you hope to conceal a theft for any length of time? ... Those women must rob you of your common sense .... Or are you made of different stuff from us?"
>
> (Balzac, 1909, 211)

This statement by the Prince goes to the heart of my argument about Balzac's implicit nativism: Balzac clearly implies that people like the Baron are *made* differently from the rest of us. The Baron possesses knowledge relating to the risks associated with his choices, as the Prince highlights, but this knowledge appears to have no traction on his behavior. This striking capacity for denial and self-deception may be a function of the detaching mental defenses alluded to above, however such behavior may also be related to the fact that the brain's reward center (a collection of dopamine rich cells known as the nucleus accumbens) is a device ruled only by bursts of dopaminergic firing. Unfortunately, under conditions of addiction, later-evolved frontal/ executive systems exhibit impaired modulation of the phylogenetically ancient seeking system (Ross, 2013, 41). The non-addict appears able to choose when or how long to enjoy the pleasures of this ancient device. However, addicts are people who are more prone to certain neuronal adaptations—in response to specific reinforcers—that permanently increase the motivational salience of cues predicting reward leading to the loss of volitional control over behavior and perseverative reward seeking (Kalivas and Volkow, 2005).

Interestingly, these neuroadaptional changes occur not only within the ancient nucleus accumbens that lies deep within the brain in the subcortex, but there are also persistent changes in excitability in more recently evolved regions in the frontal cortex that detect and respond to cues predicting rewards—regions that are critical for assigning salience and value to stimuli and for the initiation and direction of behavior. Because of the involvement of cortical regions, it is no surprise that addicts exhibit cognitive alterations—including impaired decision-making, diminished self-control, as well as poor insight into the true source of their motivation. In the tautological universe of addiction, if something exerts irresistible motivational force then it is worth any and all costs attached to its pursuit. Thus, under the influence of addiction, the addict's enduring priorities become completely upended: the subcortical tail wags the neocortical dog.

Over the course of the novel, the Baron's behavioral control, as in classic addiction, erodes, and calamities of greater degree have less and less impact upon his pursuit of sex. In addition to becoming inured to even the direst consequences of his behavior—including bankruptcy and the suicide of his uncle—prior sources of satisfaction and purpose no longer hold any interest for the Baron. Balzac's depiction is, here again, remarkably consistent with what science can now explain about persistent changes seen in established addicts: not only are the reward/ motivational pathways permanently sensitized to cues predicting addictive rewards, but in the absence of such cues the reward system is under-active, leading to reduced interest in natural rewards like social affiliation and various other natural reinforcers that non-addicts seek out and enjoy. This explains why addicts neglect prior sources of pleasure like relationships or work and why their behavior appears so aberrant to those around them. Consistent with this, the Baron's family and associates come to view him not as a dandy but as a person suffering from a bona fide disease: he is pitied as much as pilloried. An officer who arrests him for indecency, declares that he is as much in need of a physician as a sympathetic magistrate: "I respect an inveterate passion, as a doctor respects an inveterate complaint .... Such passions as these are like the cholera" (Balzac, 1909,183). In this and other ways, Balzac emphatically distinguishes Hulot from the other male characters who display various degrees of obsessive infatuation. The Baron's behavior has little to do with romantic or erotic love or even the vanities associated with feeling loved: he has become an animated puppet. Balzac appears to have swallowed our contemporary addiction-as-disease model hook, line, and sinker!

Indeed, as the story proceeds, the Baron's increasingly bizarre behavior is driven less by the pursuit of pleasure and more by a frantic need to alleviate the negative emotional state induced when a woman periodically pulls away from him. Such behavior is evocative of what contemporary addiction researchers describe as the progression from early to late stage addiction "where impulsivity (entrained by positive reinforcement) often dominates at the early stages and compulsivity (driven by the negative reinforcement characteristic of states of withdrawal) dominates at terminal stages" (Koob, 2009, 18).

Those skeptical of treating Hulot's behavior as a legitimate sub-type of addiction might object: clearly these behavioral disorders must be distinct from classical addiction as they lack the physiological tolerance and withdrawal found in intoxicant addiction. However, the phenomena of tolerance and withdrawal alone cannot explain addiction. We know this to be true because the addict's initial overuse of substances occurs in advance of significant tolerance and withdrawal and simple re-exposure to reward-related cues can cause relapse even after years of sustained abstinence—that is, not under the influence of physiological withdrawal (Hymen, 2005). In other words, tolerance and withdrawal are simply a function of the unique biological properties of intoxicants, but these unique physiological properties in no way provide evidence against the commonalities underlying both drug and behavioral addictions. Thus, if one believes, as Balzac clearly did, that the Baron is *made* differently in much the same way as the classical addict, then the common behavioral proclivities seen in both forms of addiction represent more than resemblance: their apparent congruence likely indicates a shared biological substrate (specified by common genetic antecedents) and so both sub-types of addiction likely represent a unitary natural kind.

Consistent with Balzac's fascination with addiction, Balzac endows Madame Marneffe—who can be said to inhabit the libidinal sinkhole of the story—with a preternatural grasp of the rules that govern pathological seeking. Marneffe is depicted as villainous but also as a prodigy of sorts. She is not only Balzac's avenging angel (overturning conventional morality and redistributing wealth), she is a natural genius, an artist of ensnarement and manipulation. Like the gaming industry, she understands the power of irregular schedules of reinforcement—namely, irregular schedules of reward induce far more powerful responses than fixed or predictable schedules of reinforcement. Thus Marneffe's calculated restraint and fickleness are shown to amplify desire. In Balzac's world, coyness and chastity (in Marneffe's case obviously feigned) prove to be the most potent of aphrodisiacs:

> Madame Marneffe, twenty-three years of age, a pure and bashful middle-class wife, a blossom in the Rue du Doyenne, could know nothing of the depravity... of harlotry ...; he had never known the charm of recalcitrant virtue, and the coy Valerie made him enjoy it to the utmost.
>
> (Balzac, 1909, 67)

Balzac explores this theme *ad nauseam*; he clearly understood that seeking-behavior responds most robustly to scarcity, to unpredictability—the unexpected thrill of virtue blossoming into depravity. This should come as no surprise to anyone with a passing familiarity with the stratagems of the pornographer or the gaming industry: namely, a critical aspect of what motivates animals to pursue reward is uncertainty. This is true because dopaminergic uptake in the brain's reward pathways occurs in response to environmental cues that the seeking system has become conditioned to associate with unpredictable or larger than expected rewards (Ross, 2013, 41).

In other words, animals do not need to attend and learn predictive cues in an environment where things are theirs for the taking; rather the seeking system has been attuned, by natural selection, to respond to environmental cues that matter—matter in so far as they predict a future reward that has a degree of temporal and quantitative unpredictability that requires (and justifies) the attentional, learning, and motivational resources devoted to it. Given these parameters of the seeking system, one can appreciate that the gambling environment and the bedrooms of nineteenth-century Paris share many characteristics.

This feature of reward-related learning also explains why certain categories of reward are more likely to induce behavioral addiction.

Because gambling so closely adheres to the pattern of intermittent, unpredictable, and greater than expected reward, it is no surprise that it is one of the most common and intractable behavioral addictions. For similar reasons, true sex addiction may be fairly uncommon as the cues and rewards are usually too temporally separated to entrain the archaic reward-encoding brain regions in the manner that gambling or drug use can (Ross, 2013, 45).

If Marneffe is the psychological savant that channels Balzac's insights into the rules of the seeking system, then the Baron's wife Adeline is surely one of the most clueless figures in the history of fiction. In terms of narrative convention, she serves as a foil to Hulot—embodying the virtues of purity, filial devotion, and compassion. However, Balzac adds an absurdist dimension to this character by making her ridiculously forgiving of the Baron. Her tolerance of her husband's outrages, as well as the absence of appropriate anger towards him, makes her a co-conspirator in the financial ruin of their family. One cannot help but view her as a sort of pathological empath, virtually incapable of asserting self-interest if doing so might injure anyone. Although in contrast to the impulse-ridden Baron she is a paragon of loyalty and sacrifice, her behavioral proclivities are shown to be almost as maladaptive and inflexible as the Baron's. Hence, Balzac renders Adeline and the Baron as Janus-like figures—she may be better behaved then the Baron, but her choices are just as inexorably determined. It may come as no surprise that Balzac translated Spinoza as a youth and indeed he appeared to share the philosopher's determinism and his skepticism concerning free will; accordingly Adeline, as much as the Baron, is shown to be a victim of her own fixed proclivities. Thus for Balzac, the crypto-nativist, the principle of mental life is stasis—the protagonists' choices may vary according to the changing contingencies of the moment but these choices are always true to the parametric settings that control choice for the type in question.

Despite the fact that Adeline is largely a two-dimensional creation, her brief scenes with Crevel are pivotal moments that frame the entire enterprise. In the first scene, she is offered the opportunity to prostitute herself to Crevel to save her family. She is as incapable of submitting to Crevel as the Baron is of controlling his lust. In a bookend scene near the close of the novel, the family's situation has become so dire that she now reverses her prior decision and offers herself to Crevel. In this moment, virtue and vice are held in blasphemous equilibrium as selflessness masquerades as vice and adultery becomes an act of filial devotion. But despite her good intentions, Adeline cannot be who she is not: she applies rouge and wears a dress with a low neck line but she cannot become a seductress at will. If there were any chance that Crevel might have accepted her gambit, this is ruined when Adeline dissolves into unsexy tears. He gently rebuffs her, explaining that she could never emulate Valerie Marneffe: "Valerie is a masterpiece in her way .... Twenty-five years of virtue ... are like a badly treated disease. And your virtue has grown very moldy, my dear child" (Balzac, 1909, 198).

Balzac gratuitously heaped humiliation upon poor Adeline. Perhaps this constituted a form of revenge for Balzac the private citizen—against the decent, parsimonious citizens who clucked their tongues at the great man as he was dragged away by his creditors to jail. But Adeline probably served a number of functions for Balzac, including providing a vehicle for exploring a theme that clearly intrigued him: namely, the complex relationship between vice and virtue. When taken to extremes (as in the cases of the Baron and Adeline), vice and virtue begin to resemble each other. Abstemiousness, it seems, has its own pathologies. As Howard Rachlin, author of *The Science of Self-Control*, has argued, abstemiousness often represents a "revolt against indulgence," a line in the sand that helps one "devise an overly simple solution to [the] complex problem [of temptation]" (Rachlin, 2000, 129). And yet for many addicts, survival often depends upon hewing to this binary solution. This explains why the highly structured rules and rituals of the AA movement are so helpful for many substance abusers. Addicts are by their natures suited to an all-or-nothing universe—be it a bender or cold hip-baths. They need bright lines, restrictions in their freedom of choice. Hence, they often replace drinking with overwork or compulsive exercise. As we shall now discuss, Balzac had such deep insight into these problems because he himself shuttled between extremes of overindulgence and workaholism.

Reading about Balzac's life one encounters a number of predictable accolades: larger-than-life, Promethean, a person of immense contradictions. But perhaps more accurate descriptors might be impulse-ridden, exploitative, child-like. Some, like biographer Graham Robb, have argued that his chronic problems with reckless spending and debt may have actually helped to sire the birth of the modern novel: without his repeated business failures and compulsive shopping, he would not have been compelled to attempt to write his way out of debt by penning over one hundred novels (Robb, 1994, 117).

What specifically do we know about Balzac and addictive behavior? He left a lengthy trail of financial and legal entanglements that serve to document a pattern of ill-conceived, impulsive, financial speculation and compulsive spending that left him chronically in debt. And there was no shortage of angry associates and creditors eager to go on record about the man. Providing further clues are the events and characters in his novels that were drawn from his own life.

Beginning with our primary concern—genetic influences—it appears likely that the Baron was modeled after Balzac's father, Bernard-François. Not only was the elder Balzac engaged, like the Baron, in the business of supplying military provisions to the French army, he shared Hulot's interest in the travails faced by young women compromised by circumstance. He, in fact, wrote a medical-sociological treatise entitled *The Scandalous Disorders Caused by Young Girls Being Betrayed and Abandoned in Utter Destitution* (Maurois, 1965). But his efforts on behalf of the vulnerable were interrupted, on at least one occasion, by the inconvenient pregnancy of a young woman—and this when he was 80. By then, Balzac's long-suffering mother was adept at covering up scandals and deflecting extortion threats, just as Adeline repeatedly had to rescue the Baron from himself (Maurois, 1965, 127).

Although Balzac had numerous liaisons, he was no Hulot nor even a Bernard-François. Rather it was his spending behavior—specifically compulsive collecting of antiques and paintings—that bore the hallmarks of serious behavioral dyscontrol. Interestingly, when examining the larger pattern of his behavior, it becomes apparent that his shopping was actually a form of gambling. He combed print and furniture shops for that rare, overlooked masterpiece; clearly what drove his economic activity was the pursuit of a jackpot. This pattern was apparent not only in his collecting behavior, but also in his speculative investments as well as his romantic life. He was always angling for a windfall—be it a rare find at an auction or a literary groupie with disposable income.

The late 1820s marked a period of risky financial speculation that piled up debts that would plague him throughout his life. After realizing publishers were making good money on his risqué early novels, he founded a publishing firm; when that failed, he opened a printing business that also went bust. In the 1830s he again tried to escape literary piece-work by trying his hand at magazine publishing. These efforts failed as well, augmenting his debts. Despite eventually gaining a significant literary reputation, his out-of-control behavior continued unabated. He often plagiarized and broke contracts to juice productivity and to stay ahead of creditors (Robb, 1994, 237). Perhaps Balzac's most outrageous gambit was his 1838 trip to the then primitive island of Sardinia where, based on a tip from an Italian businessman, he hoped to revive an abandoned silver mine. He was double-crossed and was lucky to make it home alive. Half-baked schemes to buy plantations and gold mines in South America fortunately never materialized.

The fortune hunting continued throughout his life, but largely took the form of maniacal collecting which, as noted, was clearly a form of compulsive speculation thinly disguised as aesthetic pursuit. His friends, family, and biographers were not fooled. Publisher Auguste Lepoitevin wryly commented that Balzac had apparently given up on becoming a novelist and, instead, aspired to being a "furniture sales-man" (Robb, 1994, 367).

Balzac's other "cousin" novel—*Cousin Pons*—is nothing less than a paean to the pleasures of collecting. Pons, a transparent Balzac stand-in, is a composer whose true passion is scavenging the countryside for overlooked paintings, prints, and antiques. But collecting is, above all, the opportunity to get something for nothing—or almost nothing—a forgotten masterpiece for 200 francs. Pons's behavior is, in fact, explicitly described as resembling that of a hashish or opium addict. And he confesses that if he had to choose between greatness as a musician or his curios, he would have always chosen his cherished collection (Balzac, 1968, 25).

Like many at the mercy of their vices, Balzac showed ambivalence toward his own behavior—an ambivalence that manifested in his often contradictory political stances. His literary realism, with its focus on the plight of the disenfranchised and the excesses of the commercial classes, made him a hero to Marx and Engels. But in actuality he was a reactionary who favored an enlightened ruling class as a check against a corrupt society and a grasping middle class. Of course Balzac, the shopaholic who aspired to owning 365 waistcoats, ironi-cally had more in common with Crevel, the ambitious striver of mod-est origins, than he would have cared to admit (Robb, 1994, 262).

It is clear, then, that Balzac's interest in addictive behavior was any-thing but accidental. His own lifelong struggles with self-control gave him uncanny insight into the habits of the seeking system. Although the impulse-ridden type has been memorialized in literature for centu-ries, Balzac's explicit portrayal of the dimensions undergirding behav-ioral addiction was remarkably perceptive. Apart from understanding the disorder from the ground up, he was also clearly aware of the ori-gins of his difficulties in his paternal bloodline. Balzac, like his father and his literary stand-in—the Baron—were in fact *made* differently. Although he could not take time away from settling his debts by churn-ing out novels, there is little doubt that if he had had the opportunity to dabble in philosophy, he would have landed squarely in the nativist camp.

References

Balzac, H., (1909). *Cousin Betty* (J. Waring, Trans.). Kindle public domain edition.

Balzac, H. (1968). *Cousin Pons* (H. Hunt, Trans.). New York: Penguin Books.

Blum, K., Bailey, J., Gonzales, A., et al (2011). Neuro-genetics of reward deficiency syndrome (RDS) as the root cause of 'addiction transfer': a new phenomenon common after bariatric surgery. *Journal of Genetic Syndromes & Gene Therapy*, 2012, 1, 5157–7412.

Green, L., Myerson, J. (2004). A discounting framework for choice with delayed and probabilistic rewards. *Psychological Bulletin*, 130, 5, 769–783.

Hymen, S. (2005). Addiction: a disease of memory and learning. *The American Journal of Psychiatry*, 162, 8, 1414–1422.

Kalivas, P., Volkow, N. (2005). The neural basis of addiction: a pathology of motivation and choice. *The American Journal of Psychiatry*, 162, 8, 1403–1413.

Koob, G. (2009). Neurobiological substrates for the dark side of compulsivity in addiction. *Neuropharmacology*, 56(Suppl 1), 18–31.

Maurois, A. (1965). *Prometheus: the life of Balzac* (N. Denny, Trans.). New York: Carroll and Graf.

Olsen, C. (2011). Natural rewards, neuroplasticity, and non-drug addictions. *Neuropharmacology*, 61, 7, 1109–1122.

Rachlin, H. (2000). *The science of self-control*. Cambridge: Harvard University Press.

Robb, G. (1994). *Balzac: a biography*. New York: W. W. Norton & Company.

Ross, D. (2013). The picoeconomics of gambling addiction and supporting neural mechanisms. In Levy, N. (Ed), *Addiction and self-control: perspectives from philosophy, psychology, and neuroscience*. New York: Oxford University Press.

5 Seymour Kety and American Lysenkoism

The novelist and entomologist, Vladimir Nabokov, was known for his fierce opposition to the two most influential intellectual movements of the early twentieth century: Marxism and psychoanalysis. While he humorously lampooned Freudians in his novels, he harbored more serious animus toward the communists who drove his family into exile and indirectly caused the assassination of his father. He sometimes remarked that however ridiculous many of Freud's ideas were, psychoanalysts at least did not go around butchering people.

Nonetheless, Russian communism and psychoanalysis did have one thing in common: an irrational hostility toward the science of genetics. The chief manifestation of this in the Soviet Union was known as Lysenkoism—a pseudoscientific movement that made every effort to counter and suppress Mendelian genetics. Endorsed by Stalin himself, this movement led to the elevation of Trofim Lysenko, an obscure Ukrainian agrarian researcher with little formal scientific training, to the pinnacle of the Soviet scientific establishment. Lysenko claimed that he could modify the characteristics of certain plants within a generation by exposing them to variable environmental conditions. Such a claim violated the accepted laws of inheritance—namely, that the units of heredity are sequestered within the germ cells and therefore not directly modifiable by the environment. His misguided attempts to modify winter wheat actually worsened famines. He also had a hand in the persecution of Soviet geneticists who refused to fall in line. Because of this, Lysenkoism is generally regarded as one of histories most notorious pseudoscientific movements.

To understand how an obscure, provincial agronomist became the director of the Institute for Genetics at the USSR Academy of Science, one must appreciate the source of Soviet hostility toward the field in the decades after the revolution. Marxist idealism presupposed that reversing the malignant conditions of capitalism could fundamentally

DOI: 10.4324/b23263-8

alter established social and economic behaviors. Therefore the Bolsheviks needed to view behavior as thoroughly malleable and responsive to changing social and economic conditions. But even in the early decades of the century, the majority of geneticists had already concluded that evolutionary transformations were gradual, unpredictable, and chiefly emerged from existing genetic variation and chance mutations. Such inconvenient facts had to be ignored or opposed and so Lysenko eschewed many of the established principles of Mendelian genetics and evolutionary biology. Instead, he promulgated outdated ideas—such as the inheritance of acquired characteristics—consistent with the directed, rapid transformation of plants, animals, and humans under variable conditions. It is no surprise, then, that when Lysenko claimed he could increase grain yields in northern regions by simply exposing seeds to cold water, this radical environmentalism was enthusiastically embraced by Stalin: here was both a solution to recurrent famine and an indictment of the "fetishes of old bourgeois science" (Soyfer, 2003, 4).

Unfortunately for legitimate Soviet geneticists, Lysenko may have been a weak scientist but he was an able and ruthless apparatchik. During his tenure, the number of biologists persecuted at the All-Union Institute of Plant Breeding exceeded the number of scientists in all fields that were imprisoned, killed, or driven into exile under the Nazi regime (Kolchinsky et al., 2017). It took decades for Lysenko to be fully discredited, and many talented scientists, such as the great Russian geneticist Nikolai Vavilov, ended up in prison or dead.

As in the saga of the Soviet misuse of science, American Lysenkoism (the anti-Mendelian strain in American psychiatry) had its heroes and villains. Seymour Kety, perhaps the most important of the heroes, was a gifted physician-researcher who was troubled by the effort of psychoanalysts, even in the late to middle twentieth century, to use their ascendant cultural and academic authority to oppose genetic theories of mental illness. Like Lysenko, these analysts perceived progress in genetics as a threat to their environmentalist agendum. And so not surprisingly, Kety, who was interested in the genetic basis of mental illness, faced opposition from many in academic psychiatry. Fortunately he was able to garner sufficient support and eventually completed landmark studies on the heritability of a major mental illness—schizophrenia. These studies struck a decisive blow against environmental theories of mental illness and would help end the hegemony that psychoanalysis had maintained over American academic psychiatry.

The belief that schizophrenia was a brain disease with substantial heritability was, of course, nothing new. Many physicians had long noted that psychosis clustered in families and that it exhibited the features of other physical illnesses such as stereotypical clinical features and a characteristic course. Even in the nineteenth century a number of psychiatrists, such as Emile Krapelin, knew that psychosis was a syndromal disease state and showed little patience for psychoanalytic theories of mental illness.

Nonetheless, given the limited effective treatments available until later in the twentieth century, one can appreciate how the Freudians came to see themselves as compassionate humanists protecting patients from the crude and often dangerous interventions of the early to mid-twentieth century—such as insulin therapy and lobotomy. Many analysts believed that they alone possessed the patience, compassion, and training to understand a disorder in terms of the patient's discourse and experience. Despite most being well-meaning, sympathy for these analytic crusaders should be tempered by the fact that the tales they spun often explicitly blamed the disorder on the patient's parents—certainly an overweening offense in light of continued and stubborn resistance to empirically based findings well into the second half of the twentieth century.

Like Krapelin decades earlier, a number of biologically oriented researchers in the United States in the mid-twentieth century were dismayed by the influence of the psychoanalytic community. According to Kety's own account of ideological forces at play in the 1960s, the combination of community psychiatry's emphasis on societal ills as a critical factor in mental illness and psychoanalysts' focus on inadequate, pathogenic child-rearing led to the gutting of funding for biological research at The National Institute of Mental Health (NIMH) in the 1960s. At one point, Kety left his post at the NIMH to seek a platform for his research in academic departments less hostile to genetics and neurobiology (Squire, 1996). As we shall see, even after Kety proved the substantial heritability of schizophrenia, a number of prominent academic psychoanalysts, including Yale's Theodore Lidz, could not accept the findings—they were apparently so inoculated against evidence by ideology.

As noted, there were many researchers and clinicians, prior to Kety, who had observed the clustering of schizophrenic-like symptoms in families. A few early twin studies showed concordance rates of the disease well over 50% in monozygotic twins, but as these twins were generally raised in the same household, environmental influence could not be excluded. As Kety himself acknowledged, many things clustered in families that likely did not have a significant genetic basis: malnutrition, level of education, chosen profession, to name a few examples. Teasing apart environmental and genetic contributions required experiments that could control for degree of relatedness and shared environmental factors. In the modern era, adoption studies using cohorts of monozygotic twins raised together and those raised apart became the gold standard for deriving measures of the heritability of psychological disorders as this method helps disentangle environmental and genetic influences. However, the low incidence of schizophrenia in the general population made adoption studies of schizophrenic twins impractical.

Schizophrenic twins might have been in short supply, but a young psychiatrist working in a state hospital in California in the 1960s, Leonard Heston, discovered that there was an abundance, for obvious reasons, of adopted-away children of schizophrenic mothers. Like many foundlings at this time, these children were raised in a variety of settings but almost all had no contact with their biological kin. Heston's design was simple: he compared the rate of schizophrenia in these offspring of schizophrenic mothers with the rate of the disease in matched controls given up for adoption by non-schizophrenic mothers.

The results were highly significant: of the 47 adoptees born to schizophrenic mothers, 16% developed the illness while none of the 50 control subjects became ill (Heston, 1966, 825). The study was quite rigorous with blind and reliability-tested diagnostic procedures and with carefully matched control groups. Furthermore, the rate of illness in the offspring of schizophrenic mothers in this study was comparable or slightly higher than the rates found in large population studies of first degree relatives of schizophrenics. Thus, Heston's study showed that removing a child from a schizophrenic mother did nothing to reduce that child's risk of developing the disorder. This, of course, suggested a very strong genetic component and a minimal to nil home environment contribution to the disease. Buried in the data was a finding that would be amplified later by Kety's much larger adoption studies: the presence of odd beliefs, social isolation, and low function in a significant percentage of the adopted-away children of schizophrenic mothers who did not warrant the formal diagnosis of schizophrenia. These traits suggested that the genetics of schizophrenia might be quite complex and that variants of the condition might exist in a partially submerged form in family members of those with the full-blown disorder.

Although Heston's paper appeared to confirm what was patently obvious to many, it lacked the size and scope of a definitive study—the type of study that would loosen the hold that Freudians still maintained over American psychiatry in the 1960s and 1970s. In many ways, Seymour Kety was the perfect person to fire the shot over the bow of the *USS Vienna*. He had no obvious axe to grind as he was a basic scientist who never even trained in clinical psychiatry. Some Freudians used this to discredit him but it actually gave him the valuable perspective of an outsider. Furthermore, his scientific achievements, prior to tackling the genetics of schizophrenia, showed a high level of rigor and imagination. Kety had refined a very early method of functional brain imaging—having performed the first regional blood flow studies of the mammalian brain using radioisotopes. These imaging techniques were the forerunners of the functional MRI scans that now yield highly detailed maps of regional brain activity.

Despite his accomplishments and technical prowess in the laboratory, Kety was drawn to population genetics partly because of his frustration with the ideological-driven resistance to science that he encountered in behavioral medicine. The analyst Ted Lidz proved to be a particularly persistent, nettlesome antagonist to Kety. As Dolnick details in his superb account of psychoanalytic resistance to science, *Madness on the Couch: Blaming the Victim in the Heyday of Psychoanalysis*, Lidz was responsible for peddling some of the most egregious theories that explicitly blamed parents for their children's schizophrenia (Dolnick, 1998).

Reading some of Lidz's papers on the environmental causes of schizophrenia, one is struck by a virtual catalogue of logical fallacies. These are many of the same fallacies that Krapelin accused Freud of in the early twentieth century: the extrapolation of theories of causation based upon anecdotal case histories; the assumption that the mental content brought forth by the patient was helpful for understanding the etiology of the disorder; and, finally, the putting forth of conjecture as established fact. Lidz argued that the disordered associations and scattered thinking of schizophrenia did not represent a brain disorder but a reaction to "the severely disturbed thought ... processes of the parents" (Lidz, 1968, 177). According to this theory, conflict between the parents and/or their inability to model normal emotional, linguistic, and interpersonal behavior led the child to "distort perceptions of his internalized version of his world to gain some spurious resolution of untenable conflicts" (Lidz, 1968, 176). In this way, Lidz argued for the environmental induction of the illness through an experiential "transmission of irrationality" from parent to child. As was the case with Lysenko, empirical genetic research threatened Lidz's radical environmentalism which presupposed that external conditions could directly transform complex cognitive and emotional processing systems.

Lidz's model, like Lysenko's effort to transform winter wheat, runs according to a facile, cause-and-effect logic. But in Lidz's case, there is an added helping of baroque contrivance. Let us try, however awkwardly, to follow the outlines of Lidz's over-freighted theory: the child is faced with a warped, un-metabolizable family scene created by the parents' impaired ways of relating; the child's attempt to assemble a tenable reality produces a parallax correction that distorts normal reality and disrupts the basic perceptual, cognitive, and linguistic functions of the child's mind. This process induces not only psychosis but the cognitive unraveling characteristic of schizophrenia. Through it all, there is a unitary view of mental function as an all-purpose, conflict-reducing mechanism. As was seen in the Freudian schematic, all symptoms are viewed as displaying instrumentality—operating like protective mental tropisms that fulfill a purpose according to an absurdly stereotyped logic. Such thinking entails the most obvious metaphysical error—that the natural world is organized according to rules that reflect human-like rationality. In addition to these absurdly speculative theories, Lidz would also question the results of decades of painstaking, careful work by Kety and colleagues—even into the final decades of the century.

The research conducted over many years by Kety and his collaborators represented one of the largest and most ambitious investigations of the inheritance of a major psychiatric disorder ever conducted. Like Heston, Kety used adoption to help disentangle genetic from environmental influences. Taking advantage of the excellent adoption and mental health records maintained in Denmark, Kety first conducted a small-scale study in Copenhagen and then a larger national Danish study of schizophrenics who had been adopted-away soon after birth. If schizophrenia were principally a genetic-based disorder, one would expect the adoptive families of these index cases to be relatively free of the disorder, while the biological relatives would show elevated rates of the disease. This is exactly what was found. In both the provincial and national studies, chronic schizophrenia was found in approximately 5% of the biological relatives of the schizophrenic adoptees but in less then 1% of adoptees in the control samples—reflecting the baseline rate of the disease in the general population (Kety et al., 1994, 449). Furthermore, there were no cases of chronic schizophrenia in the adoptive relatives of the schizophrenic adoptees. This confirmed that the clustering of the disease within families that had been noted since the nineteenth century had to be the result of genetic factors.

The study also helped to confirm and refine another aspect of the familial transmission of the disorder—an aspect that had been observed by Heston as well as psychiatrists in the nineteenth century: namely, that a milder, less disabling form of the disease was often found among blood relatives of schizophrenics. The Swiss psychiatrist Eugen Bleuler described this milder variant, what he had termed "latent schizophrenia," in his classic monograph *Dementia Praecox, or the Group of Schizophrenias* (Bleuler, 1911). Bleuler noted that the relatives of his patients often exhibited many of the same behavioral traits seen in chronic schizophrenia minus the florid psychotic symptoms of hallucinations and delusions. These traits—such as suspiciousness, flattened affect, social withdrawal, and odd beliefs—constituted, he believed, a milder or marginal variant within a schizophrenic spectrum. Physicians in many specialties had long observed that disease type but not that disease severity tends to run in families, so Bleuler was simply extending a pattern seen in a number of medical illnesses. Kety, therefore, was not attempting an end-run around weak data (as Lidz would later contend) when he included data on latent schizophrenia in his study; rather, Kety used his data to help clarify the boundaries of the previously described schizophrenic spectrum. Kety's data also confirmed Bleuler's suspicion that the milder variant was in fact the most common form of the disorder found in family members. (The mild variant was found in 10.8% of biological relatives of schizophrenic adoptees versus 1.7% of the biological relatives of control adoptees in the national sample.)

Perhaps to counter anticipated criticism from the psychoanalysts—who viewed mental illness less in terms of discrete diagnosable diseases and more as a continuum running from normal to neurotic to psychotic—Kety looked at the prevalences of other mental disorders in these families. He found that no other diagnoses outside the schizophrenic spectrum were found at higher frequencies in the relatives of either index or control adoptees. "This suggests," Kety and colleagues concluded, "that [the] increased risk for psychiatric illness in the biological relatives of schizophrenic individuals is limited to the diagnoses of schizophrenia and latent schizophrenia, rather than a risk of psychopathic features in general" (Kety et al., 1994, 450).

Despite the overwhelming evidence that the familial clustering of the disease was an expression of shared genetic factors *and* that sharing a family environment contributed absolutely no additional risk, Kety was careful not to overstate his case: he acknowledged that his findings did not exclude the possibility that non-genetic factors might influence the age of onset or the course of the illness. Despite Kety's restraint, Lidz raised a number of pointed objections to the Danish studies in a series of letters to a leading journal (Lidz and Blatt, 1983). For example, Lidz assumed that Kety had compared the rates of schizophrenia in biological relatives of those with the disease against biological relatives of control adoptees because a direct comparison between blood and adoptive family members had failed to reach significance. However, as Kety explained in his response to Lidz, although the comparison held in both instances, adoption studies are more rigorous if they utilize control groups that compare like with like (i.e., blood relatives of schizophrenics to blood relatives of controls, adoptive relatives of schizophrenics to adoptive relatives of controls). In this way, one may control for potential confounding variables such as characteristics unique to parents who give up or adopt children. Lidz also questioned Kety's use of half-siblings in his data, though including half-siblings actually adds credence to adoption studies because paternal half-siblings do not share intra-uterine environments—a potential non-genetic source of variance. Additionally, Lidz objected to Kety's inclusion of an "indefinite category"—latent schizophrenia— in his study. However, as Kety also explained:

> the very vagueness and lack of sharp boundaries would, if anything, have diminished the likelihood of our finding significant differences blindly between the index and control biological relatives. All the more remarkable, then, was the significance of the difference that resulted in the case of latent … schizophrenia.
>
> (Kety, 1983, 964)

Others would go on to independently confirm the reliability of Kety's provisional diagnosis of latent schizophrenia; this presumed mild subtype of schizophrenia would later be included in DSM-III as schizotypical personality disorder. Perhaps most importantly, this demonstrated that population genetic studies could provide an empirical basis for assessing the validity of psychiatric diagnosis by establishing genetic antecedents of a disorder (Kety et al., 1994, 454).

Ted Lidz could probably not forgive Kety for proving that the so-called schizophrenigenic family was nothing but a psychoanalytic myth. Lidz was certainly no Lysenko. His criticisms of Kety likely provoked eye-rolls rather than trepidation. Furthermore, Lidz had plenty of company 50 years ago and many to this day still defend psychoanalysts from that era—citing the limited knowledge of genetics and neuroscience at the time. However, there were also many, like Seymour Kety, who objected to palpable nonsense and made significant progress with the tools available. So while one can perhaps understand how many clinicians were swept up by the power of Freud's genius (and applied the best of his ideas to refining our understanding of mental defenses, for example), it remains troubling that some continued, despite increasing evidence, to push unsubstantiated environmental theories of schizophrenia. Perhaps Nabokov went easy on these mid-century zealots after all.

References

Bleuler, E. (1911). *Dementia Praecox, or the group of schizophrenias*. New York: International Universities Press.

Dolnick, E. (1998). *Madness on the couch: blaming the victim in the heyday of psychoanalysis*. New York: Simon & Schuster.

Heston, L.L. (1966). Psychiatric disorders in foster home reared children of schizophrenic mothers. *British Journal of Psychiatry*, 112, 819–825.

Kety, S. (1983). Letter. *The American Journal of Psychiatry*, 140, 7, 964.

Kety, S., Wender, P., Jacobsen, B., et al. (1994). Mental Illness in the biological and adoptive relatives of schizophrenic adoptees. *Archives of General Psychiatry*, 51, 442–455.

Kolchinsky, E.I., Kutshera, U., Hossfeld, U., Levit, G.S., (2017). Russia's new Lysenkoism. *Current Biology*, 27, 19, 1–29.

Lidz, T. (1968). The family, language, and the transmission of schizophrenia. *Journal of Psychiatric Research*, 6, 1, 175–184.

Lidz, T., Blatt, S. (1983). Letter. *The American Journal of Psychiatry*, 140, 7, 963.

Soyfer, V. (2003). Tragic history of the VII International Congress of Genetics. *Genetics*, 165, 1–9.

Squire, L. (1996). *The history of neuroscience in autobiography*, vol. 1. Washington, DC: The Society for Neuroscience, 382–413.

Part III
Reconnoitering Innateness

6   Innateness Wars
A Darwinian Aside

While most of this volume is concerned with the heritable influences that contribute to the behavioral variance found among individuals, in this chapter I assume an evolutionary perspective and so focus at the level of the species. Assuming this perspective, the following questions will be discussed. How can one lay claim to a reasonable, contemporary version of nativism that acknowledges innateness to be a relative property as almost all traits are influenced by extrinsic conditions? How can one discern the outlines of innate proclivities in a species such as ours that is characterized by complex states of awareness and high degrees of learning that would appear to swamp or obscure the outlines of innate mental structure?

Speculation on such questions date back to the Socratic dialogue on the innate mathematical sense displayed by an untutored, illiterate slave and Aristotle's commentary on brood parasitism—the instinct that leads certain birds to lay their eggs in the nests of other species. But after centuries of nativist/empiricist debate, a number of philosophers of science have recently argued that, given what we now know about the pervasiveness of environmental modulation of genetic expression, and how little we know about the "pre-specified contents ... and procedures of the mind," the concept of innateness, at least as it applies to *Homo sapiens*, has outlived its usefulness (Simpson et al., 2005, 5).

One philosopher of science, Paul Griffiths, makes a strong case that the concept is "irretrievably confused" as it "conflates a number of independent ... properties" such as unlearned, developmentally fixed, species-typical, the product of natural selection, present since birth, and so on (Griffiths, 2002, 71). But issues of conceptual cogency aside, it stands to reason that a concept touching upon the knotty problem of human nature would be attended by enduring controversy. And indeed, apart from the inexactness of such a vague and diffuse term,

DOI: 10.4324/b23263-10

there have been frequent and nasty partisan struggles in evolutionary biology over various aspects of innateness in the almost 150 years since Darwin's death. In this chapter, various dimensions of innateness will be explored by discussing such controversies—the hope being that there may be aspects of innateness that prove salvageable. Of course in the end, innateness encompasses properties that tend to be relative: traits tend to be partly heritable, may appear at variable points during development, involve relative degrees of priming or learning. Nonetheless, the existence of fuzzy boundaries does not necessarily invalidate credible distinctions.

The controversies under consideration center on scientists whose work on innate aspects of behavior provoked both scientific and political colloquy: Konrad Lorenz, the co-founder of ethology was criticized for his right-leaning politics and his flawed theory of instincts; E. O. Wilson, who sired the sociobiology movement, was a victim of the political turmoil in American academia in the 1970s; finally, Noam Chomsky, self-described anarchist/syndicalist and founder of modern linguistics, is important to any discussion of nativism. In this case, the focus will be on his more recent ideas on the nature of the language faculty which have attracted a fair degree of criticism.

The first controversy centers on the unlearned aspect of innateness. For Lorenz, innateness was defined by the complete independence from extrinsic influence. Innate behaviors could be released or triggered by the environment but were viewed as essentially inborn and inflexible. Because he was more comfortable sitting in a duck blind than studying genetics or breeding fruit flies, the movement known as The Modern Synthesis—which integrated Darwin's theories with Mendelian and population genetics beginning in the 1920s and 1930s—largely passed him by. Lorenz was chiefly interested in breaking down the mechanics of highly stereotyped behaviors, such as courtship displays. For him, the innate and the acquired were fundamentally distinct and neither one could ever evolve from the other. Obviously, this early conception of innateness was overly simplistic and could not stand. And so, by the mid-1950s, Lorenz's rigid dichotomy between instinctual and learned behavior would be challenged by American Daniel Lehman as well as figures from The Modern Synthesis like J. B. S. Haldane (Griffiths, 2004; Westneat and Fox, 2010). This would mark the beginning of a shift from classical ethology, with its focus on the mechanics of the most stereotypical behaviors, to modern behavioral ecology—a discipline rooted in evolutionary studies emphasizing an organism's responsiveness to environmental conditions and adaptive explanations of behavior.

As it turned out, Lorenz was an obvious target for scientists such as Haldane: as a Marxist, Haldane had natural antipathy toward the right-leaning Lorenz; secondly, Haldane, a philosophically sophisticated Darwinist, had no patience for the teleological flavor of Lorenz's ideas—his idealization of instinct, his denigration of domestication as a form of degeneracy, and his tendency to portray natural selection as a progressive process. Haldane had famously joked that teleology was for the biologist like a mistress: he never wanted to be seen in public with her yet he did not seem able to live without her. Although Darwinism did indeed deal with *optima*, these were always relative and from the perspective not of design but of differential survival and reproduction.

It was, in part, breakthroughs in research on differences in song acquisition among various bird species that stimulated Haldane's interest in learning and behavioral plasticity. His insights into the role plasticity could play as both a force and target of natural selection proved that Lorenz's narrow account of instinct was inadequate (Griffiths, 2004). Haldane's interest in the role of plasticity in evolution might have appeared cutting-edge in the 1950s but similar ideas date back to the nineteenth century. In traditional Darwinian theory, random mutation provides the material for selection—therefore changes in genotype lead and phenotype follows. However, in the 1890s, a psychologist and evolutionary theorist named James Baldwin proposed that variable behavioral phenotypic responsiveness to environmental shifts (behavioral plasticity) could actually nudge selection in an adaptive direction.

Baldwin theorized that individuals varied in terms of their behavioral plasticity—their ability to react to changes in the environment. As this can impact differential survival, behavioral plasticity could thus be a target and force in evolution. In other words, phenotype could sometimes lead and genotype follow. Baldwin explicitly rejected Lamarckism (the inheritance of acquired characteristics): the genetic potential for plastic responses to environmental shifts had to be already present in the population; all that was needed was a significant extrinsic change for latent environmental responsiveness to become manifest as changing conditions can reveal standing but hitherto cryptic genetic variation (Crispo, 2007, 2470).

It is perhaps no coincidence that Helen Spurway, Haldane's collaborator and partner, had a special interest in domestication, as it is an example of the kind of abrupt (albeit human-mediated) environmental shift that can exploit previously silent genetic variation (Griffiths, 2004; Spurway, 1955). In fact, it could be argued that domestication demonstrates a Baldwin-effect type process as a shift in environmental conditions—as well as an accelerated form of selection (i.e., artificial selection)—taps latent genetic variation for behavioral plasticity that can be exploited by breeders. The rapid behavioral diversification characteristic of canine breeds is perhaps an example of how efficient such a process can be in rapidly producing uncanny behavioral traits in an intelligent species (e.g., retrieving, pointing, herding). Nonetheless, the relative importance of phenotype-directed evolution of behavior under natural conditions remains a matter of debate to this day.

Seen in historical context, the Baldwin effect was a theory ahead of its time—awaiting the likes of Haldane and Spurway to resurrect and extend these prescient intuitions. The Neo-Darwinists' interest in Baldwin should come as no surprise because, in contrast to Lorenz, they realized that even fairly stereotyped instinctual behaviors showed a range of so-called reaction norms and so could vary under changing conditions. But while the Baldwin effect may have focused attention on adaptive aspects of plasticity, another British geneticist, Conrad Waddington, developed a more complex view of plasticity—a view that could also accommodate the "tension between the need to be sufficiently buffered from the environment vs. the ability to track and adaptively respond to [changes in the environment]" (Ghalambor et al., 2007, 396).

Waddington most thoroughly explored this tension between plasticity and developmental fixity and the conditions which might favor one over the other. His theory of genetic assimilation was a sort of reverse Baldwin effect: he predicted that under certain conditions, selective pressure would lead to a decrease in plasticity by absorbing a learned behavior into the genotype thus making it more independent of environmental influence, or "canalized" as he put it. This process, then, would convert behaviors that were originally highly responsive to environmental conditions into more fixed behaviors. It was hypothesized that "canalization" of plastic traits might be the second step whereby Baldwin's environmentally induced behaviors eventually became "assimilated" to the genome. Thus Lorenz's specious divide between the instinctual and the learned was breached from two directions as behaviors could apparently evolve toward greater plasticity or fixity.

In a classic experiment, Waddington was able to prove that an environmentally induced change could, in some instances, become inherited (Crispo, 2007, 2472). He subjected a population of fly larvae to an environmental stressor—heat shock—which caused a percentage of offspring to develop a variant wing pattern. He then selectively bred the flies that expressed this atypical wing pattern in response to heat. After many generations of culling and breeding heat-shock-induced variants, flies with the atypical wing pattern began to appear spontaneously without the application of heat shock. Waddington had demonstrated that an environmentally induced trait could become established in a pedigree! Molecular biologists have since discovered that there are regulatory proteins (called heat-shock chaperone proteins) that buffer certain genes against acute environmental perturbations. Waddington was apparently working with a stock containing strains with compromised chaperone proteins. When these were bred, a hitherto suppressed wing pattern emerged: in other words, there are regulatory proteins that function to suppress the expression of certain genes and therefore may act as buffers by maintaining well-canalized traits in the face of acute environmental shifts (Crispo, 2007, 2473).

Waddington had thus unwittingly discovered a specific mechanism that could regulate gene expression. He named this phenomenon an epigenetic effect and a new discipline was born. While the Baldwin effect portrayed plasticity as a driver of evolutionary change, Waddington had revealed mechanisms (e.g., gene suppressor mechanisms) that favored developmental fixity. Under certain conditions, there may be advantages to greater behavioral fixity and it is now understood that "well-canalized" traits likely play an important role in building systems that are sufficiently buffered against environmental perturbation. To be sure, natural selection does not necessarily favor greater plasticity, especially because many forms of plasticity (including those that depend on learning) generally entail considerable costs. As behavioral ecologist John Alcock has argued,

> learning does not produce behavioral change just for the sake of change. Instead, selection favors investment in the mechanisms underlying learning only when there is environmental unpredictability that has reproductive [or survival] relevance for individuals …. [As] learning mechanisms are costly, then we can expect learning to evolve only when there is some major counterbalancing benefit.
>
> (Alcock, 2009, 97–98)

Thus, behaviors that involve significant learning are not necessarily more adaptive—rather relative degrees of fixity and plasticity fluctuate in all species according to the specific selective pressures at work.

Research on birdsong in the 1950s lent credence to this view. Peter Marler, a student of Haldane's, discovered variation in the degree of learning involved in song acquisition in various bird species. Some species appeared to require little exposure and acquired a standard, invariable, species-typical song. In other species, adolescents required prolonged exposure to the songs of neighboring adults and developed a unique, local dialect. Haldane correctly theorized that local dialect acquisition served a territorial rather then a courtship function. This is why, he reasoned, that the canary had lost its dialect learning capacity in the 200 years since its domestication as territorial behaviors have little utility in domesticated animals (Griffiths, 2004). And so Neo-Darwinian advances drove home the point alluded to above—namely, domain-specific plasticity including complex types of learning will vary, like any trait, in response to specific selective pressures.

So then, what does this first controversy in the history of ethology teach us? An early focus by field biologists, such as Lorenz, on the most stereotyped behaviors implied a conception of innate as fixed and immune from environmental influence. Scientists such as Haldane and Waddington helped formulate a more nuanced conception of instinctual animal behavior. They did so by challenging the dichotomy between the "instinctual" and the "acquired" by demonstrating that many behavioral traits exhibit relative degrees and quite specific forms of environmental responsiveness depending upon the selective forces at hand. Indeed, the *ways* in which an animal learns can be as uniquely pre-specified and genetically based as the most fixed and invariant "instincts." As E. O. Wilson maintained, species-specific learning mechanisms could not have evolved if learning were a generalized, all-purpose process as there would be no mechanism for transmitting such changes to offspring. Therefore, as Wilson explained, "what evolves is the directedness of learning—the relative ease with which certain associations are made and acts are learned, and others bypassed even in the face of strong reinforcement" (Wilson, 2000, 156).

The broadening of the original ethological conception of instinct was only the beginning of a revival of nativism—concurrent with explosions of knowledge in linguistics, neuroscience, and genetics. As will be discussed below, sociobiology and evolutionary psychology—the black sheep offspring of ethology—arose and attempted to extend innateness to highly compound, culturally intwined human behaviors, such as propensities toward tribalism and religiosity. But before proceeding to the battle over sociobiology, a discussion of less controversial compound or second-order by-products of base, innate functions might be instructive. Consider, for example, the human capacity for deciphering linguistic symbols. The ability to read is obviously not inborn but acquired during development with the aid of all sorts of cultural innovations and so cannot be considered innate in any sense of the word. Nonetheless this capacity is likely built upon an innate, highly adept visual-parsing functionality that can access the species-specific linguistic processing regions of the human brain. It can also be selectively impaired by discrete genetic abnormalities or by localized neurological insults such as small strokes. All these facts imply that the capacity to read (although in proximal terms, a cultural invention) is built upon highly species-specific, modular systems. Additionally, if given adequate education, the facility with which children acquire this skill shows high levels of heritability in families. Thus, there can be no genes that evolved specifically for reading; it is, strictly speaking, a skill that was developed only recently and required very particular economic and historical conditions to become manifest. Nonetheless, reading in humans appears to at least qualify as a second-order derivation of innate endowments—an assemblage of primary capacities that emerged under particular historical conditions.

The same could be said about a range of behaviors that are likely second-order derivations of upstream, innate endowments. Such endowments, when incubated with an array of cultural and developmental influences, yield the compound traits and pathologies that are easily recognizable in our species such as frugality, introversion, the tendency toward addiction, or vulnerability to excessive procrastination. Of course none of these are innate in the Lorenzian sense—such as babbling in infants or fear conditioning; however, many of these derived traits, despite their compound and contingent natures, still descend reliably through pedigrees, show significant degrees of heritability, and are likely derived less from experience but more from primary behavioral and cognitive endowments.

The battle over sociobiology, sparked by the 1975 publication of E. O. Wilson's *Sociobiology: The New Synthesis*, constitutes a second nodal point in the controversy over innateness. Wilson's compendium of instinctual social behavior from the insects to the primates contained a brief chapter that proposed extending ethology to human behavior. It was this chapter that provoked a nasty backlash from leftist academics led by Marxist geneticist Richard Lewontin and made Wilson a poster child for reactionary Darwinian social theory. However, Michael Ghiselin, a quirky, little known but marvelously original biologist, may rightly be considered the co-founder of sociobiology as his discussion of Darwin's implicit sociology (in the monograph *The Triumph of the Darwinian Method*) was published six years prior to the release of Wilson's book.

The sociobiology battle focused on to what degree evolved adaptations might be relevant to complex social behavior in humans. Like the earlier debates over instinct, it was unfortunately framed in left versus right-wing political terms, despite the fact that Wilson and his colleagues were, for the most part, politically liberal or moderate. But there were many reasons for the left's negative reaction to sociobiology. First off, it threatened the prevailing social science model that emphasized the primacy of learning in human development. Secondly, it struck many as an exercise in runaway genetic determinism that reinforced existing power hierarchies and gender roles. Lastly, the theory of natural selection had parallels with free-market economic theories, some of which had indeed influenced Darwin. As philosopher of science Eliot Sober has pointed out "the Scottish economists offered a nonbiological model in which a selection process improves a population through an unintended consequence of individual optimization" (Sober, 1993, 20).

So for many on the far left moral hazard was, from the start, baked into a Darwinian sociology that they believed sought to tether culture to a genetic leash and reify human nature. It is interesting that two generations of leftists were offended by those who were seen as having fetishized instinct. Here we will not delve much further into the politics of the sociobiology debate. Instead, we will focus on how the battle over sociobiology sheds light on certain key problems that must be accommodated in a reasonable nativist account of human behavior. The first problem is the quandary that high intelligence presents— namely, to what degree does human intellection, and its product culture, swamp or overshadow the innate proclivities that lie at the root of complex social behavior. Secondly, what role do chance contingencies or "accidents of history" play in the development of species-specific behavioral traits—an issue that, in some respects, is analogous to the role environmental factors play in the development of individuals?

Turning first to the issue of historical contingency, obviously environmental contingency is part and parcel of the process of natural selection as an organism's fitness is always in relation to a particular set of extrinsic conditions. What Lewontin apparently objected to was a *relative* neglect of extrinsic, contingent factors that he believed marked Wilson's project—a project that Lewontin viewed as hell-bent on finding discrete adaptive mechanisms to explain all sorts of complex behavior. Lewontin's critique of sociobiology, therefore, focused on what he and Stephen J. Gould regarded as Wilson's overly zealous view of what natural selection could accomplish and his tendency to attribute too much to its specific action. Lewontin argued that the adaptationists, as he called them, downplayed the role of non-selectionist forces, such as chance environmental factors or so-called "accidents of history." In essence, Lewontin pejoratively identified adaptationism with optimization—overemphasizing the evolutionary utility of behaviors and traits while underestimating historical, serendipitous factors and design constraints (Lewontin, 1979).

Interestingly, Gould and Lewontin's complaints about sociobiology are reminiscent of Haldane's objections to Lorenz's idealization of instinct. But while Lorenz's grasp of the subtleties of Darwinism was weak, Wilson clearly understood that attempts to extend ethology to human behavior required an account of innateness that could accommodate the problem of historical contingency and other constraints on optimization. As Wilson's colleague Michael Ghiselin repeatedly pointed out in his monograph *The Triumph of the Darwinian Method* (1969), Darwin himself had extensively addressed chance and other non-selectionist forces in evolution: "Darwin thought that many behavioral phenomena ... resulted through accidents of history" (Ghiselin, 1969, 205), and that Darwin's argument against design in his attack on natural theology presumed that "the properties of organisms are ... explainable as the consequence of piecemeal, blind, and often, in the long run, deleterious changes" (Ghiselin, 1969, 158). Indeed, Ghiselin, whose book was published in 1969, six years before Wilson's book went to press, often sounds as if he were writing a direct rebuttal to Lewontin's future attack on Wilson. Ghiselin also cites Darwin's fourth notebook on the transmutation of species in which he foreshadowed, however vaguely, various constraints on optimization, such as those imposed by development, morphology, and as yet unknown genetic mechanisms such as epistasis: "laws probably will be discovered of correlation of parts, from laws of variation of one part effecting another" (de Beer, 1960, 167). And if there is any doubt about Darwin's conviction regarding the import of unpredictable forces in inheritance and development, one need only look to his remarks on the vagaries encountered by pigeon fanciers who, when breeding for certain traits in the carrier pigeon, discovered that differences often arose "not from, but rather in opposition to, the wish of the breeder" (Darwin, 1889, 167).

I have been citing Ghiselin on Darwin, but what about Wilson's own sense of restraint when it comes to the potential excesses of adaptationist accounts of social behavior? The leading historian of the sociobiology wars, Ullica Segerstrale, examined the issue in depth and concluded, with ample evidence, that Wilson did not need to be lectured on the influence of chance/extrinsic contingencies. Segerstrale argued that Wilson did not neglect non-adaptive, serendipitous factors in evolution as Lewontin and Gould charged. Rather his real transgression in his critics eyes, she believes, may have had more to do with encouraging speculation about the role of evolutionary adaptations in forging human social behaviors (Segerstrale, 2000, 106).

Richard Dawkins, a like-minded adaptationist and Wilson supporter, actually wrote a book, *The Extended Phenotype* (1982), which can partly be read as a rebuttal of Lewontin and Gould's charges of adaptationist excess and genetic determinism. In fact there is an entire chapter, entitled "Constraints on Perfection," which reviews the issue of historical contingency as well as other limits on optimization such as availability of genetic variation. While acknowledging that there were many popularizers of Darwinism that were guilty of simple-minded adaptationism and genetic reductionism, Dawkins argues that credible evolutionary biologists like Wilson were actually most interested in highly contingent behaviors as these demonstrated the complex influence of variable extrinsic conditions. For example, Wilson was particularly interested in aggressive instincts as these are almost always dependent upon resource availability and population density. As Dawkins underscores, "It is simply meaningless to speak of an absolute, context-free, phenotypic effect of a given gene" (Dawkins, 1982, 38).

Ironically, the problem of constraints on optimization—whether they be historical, genetic, morphological, or developmental—figured so prominently in Wilson's thinking that it has led to a recent schism with fellow adaptationists including Dawkins. To understand this schism one must revisit what Darwin believed to be one of the central puzzles of evolutionary biology—the emergence of altruistic behaviors in social species. In 1955, Haldane formulated a solution—the concept of inclusive fitness which was first hinted at in Darwin's attempt to explain the cooperative behavior observed in social insects. Inclusive fitness extends the traditional notion of fitness to include benefits that may accrue to an individual's kin. As the survival and reproduction of kin ensures the transmission of genes that the individual shares with these kin, the theory was thought to explain the origin of altruistic behavior characteristic of sterile castes of ants who labor in a colony of related sisters but leave no direct descendants. In 1964, a mathematically based account of inclusive fitness—an account Wilson credits with inspiring his decision to study prosocial behavior—was formalized by W. D. Hamilton in his paper "The Genetical Evolution of Social Behavior I, II." The inequality that Hamilton formulated (R is greater then C/B) predicts that cooperation will be favored by natural selection if the degree of relatedness (R) between cooperators exceeds the cost (C) to benefit (B) ratio (Hamilton, 1964).

This theory gained traction largely because it could explain the rise of the striking level of cooperation seen in the eusocial insects—those insects that exhibit extensive division of labor including sterile females that care for the queen's offspring. When this theory was first developed almost all known eusocial species were characterized by what is called haplodiploid sex determination in which females arise from a fertilized egg while males hatch from unfertilized eggs. In such species, sisters are more closely related to each other than parents are to their offspring. Thus, aiding the queen in producing more viable sisters maximizes the representation of the sterile nursemaids' genes in succeeding generations—just as the theory of inclusive fitness had predicted. However, in the decades since Hamilton's original paper, many non-eusocial haplodiploid species have been discovered; conversely, a number of eusocial diplodiploid species have been found. This led Wilson to question the theory of inclusive fitness and its role in the development of prosocial insect behavior. In 2010 Wilson and others published a controversial paper, entitled "The Evolution of Eusociality," in which they argued that eusociality likely owed its origins not to kin selection but to historical contingency. Chance factors, they proposed, such as abundant food sources and a defensible nest, play a crucial role in the initiation of communal living. It is this extended period of communal living (a chance occurrence) and not relatedness per se that permits certain preadaptations (for example, insects' tendency to perseverate on an available task) to foster the emergence of division of labor. And so if a critical number of sisters stick around a felicitously endowed nest, they will help the queen fulfill her destiny as a high-capacity breeding machine and a large colony will be established (Nowak et al., 2010).

Therefore, according to Wilson's revised account, one need not invoke a special type of gene-level selection (kin selection) when the phenomenon can be explained as a particular manifestation of natural selection fostered by the historical accident of protracted communal living. It were as if, after 1917, members of an extended family find themselves thrown together by an accident of history in a dingy Moscow apartment block with one bathroom and two hot plates. Having no better options, they decide to make the best of things; they remain together and thus by happenstance appear to embody the ideals and benefits of collectivism. So then, Wilson had in the end judged Hamilton's theory to be an over-simplification of the various forces that may contribute to cooperation. Haldane once joked that he would not jump into a frozen river to rescue a brother but would to save two brothers or eight cousins; but as one of Wilson's collaborators Corina

Tarnita has pointed out in a twist on Haldane's joke—would he have jumped in to save a cousin if that cousin were competing for reproductive access to one of Haldane's brother's prospective mates? In other words, Hamilton's ideas were indeed groundbreaking and interesting but his mathematical model was hopelessly simplistic (Lehrer, 2012).

Beyond the issue of the role that accidents of history play in evolution, Wilson's revised theory of eusociality also revived the debate on group selection (the theory that the relative survival of groups not just individuals could be a force in evolution): he argued that "between-colony selection shapes the life cycle and caste systems of the more advanced eusocial species" (Nowak et al., 2010, 1061). Wilson's questioning of the role that inclusive fitness played in insect eusociality—along with his defense of the presumed discredited theory of group selection—caused a backlash from many of his former allies in the sociobiology wars. This casts doubt on Lewontin's charge that Wilson was motivated by a commitment to narrow adaptationism. If anything, this schism demonstrated that Wilson was not afraid to break ranks with his closest colleagues in his search for the most rigorous explanation of the rise of prosocial instincts. And so it was beyond dispute: Wilson had never discounted the role of chance extrinsic factors but very much viewed adaptation, as Dawkins elegantly put it, as a "tangle of compromises" (Dawkins, 1982, 47).

In the end then, the sociobiology wars were perhaps more a reflection of conflicting philosophical and political perspectives than a substantive scientific dispute. In any case, there is no mistaking that the nativist camp views innateness, as well as the role of environment and historical contingency, in a different light than the anti-adaptationists. For the nativist, the environment functions more as a releaser of prefigured, or in Waddington's terminology, well-canalized outputs. Contexts may be infinitely variable and operate in unpredictable, dynamic ways but evolutionary (as well as developmental) outputs tend to flow within the contours of a delimited number of highly constrained pathways—pathways determined by an array of factors specific to the species in question. Darwin used a different metaphor to capture a similar idea:

> We are ... driven to conclude that in most cases the conditions of life play a subordinate part in causing any particular modification; like that which a spark plays, when a mass of combustibles bursts into flame—the nature of the flame depending on the combustible matter, and not on the spark.
>
> (Darwin, 1887, 282)

This may explain why adaptationists sometimes appear to downplay the role of chance or historical factors in the development of species (just as nativist geneticists may discount environmental/biographical factors in the lives of individuals). Anti-adaptationists, on the other hand, tend to view behavioral traits as cut loose from their genetic tether by what philosopher Robert Richards has called the "Lamarckian swiftness" of cultural transmission (Richards, 1987, 545). And indeed it is the problem of this excess potential (i.e., cognitive complexity and its product, culture) that presents the greatest challenge to the nativist perspective.

As we saw with the issue of historical contingency, the threads of this debate lead all the way back to Darwin who tried to reckon with the problem of intelligence—specifically the challenge of identifying innate traits in a species that exhibits high levels of cognitive complexity. However, Darwin clearly believed that *Homo sapiens* does not constitute a fundamental disjunction from the rest of the animal kingdom. To drive home this point he elevated lower species while tugging humanity down off its perch. For example, he wrote about attention span in earth worms and the practice of slave-taking in certain ant species while challenging the special status afforded the most prized human instinct—the moral sense. Writing in *The Descent of Man*, Darwin mused,

> If, for instance ... men were reared under the same conditions as hive-bees, there can hardly be a doubt that our unmarried females would, like the worker-bees, think it a sacred duty to kill their brothers, and mothers would strive to kill their fertile daughters; and no one would think of interfering. Nevertheless, the bee, or any other social animal, would gain ..., as it appears to me, some feeling of right or wrong.
>
> (Darwin, 1871, 73)

Now here is Wilson, clearly echoing both Darwin's use of black whimsy and his attempt to sever our sentimental attachment to our most cherished sentiments: "If like the termites we needed to dwell in darkness, eat each others' faeces and cannibalize the dead ... Our minds would be prone to extol such acts as beautiful and moral" (Ruse and Wilson, 1985, 52). Darwin and Wilson were both obviously employing a rhetorical device to puncture the self-serving illusion that our cooperative instincts represent anything other than species-specific adaptations. We may, unlike the termite, feel good when we do good, but as Ghiselin has pointed out, [however much Darwin realized that] sentiments ... [often] come into play ... their role [must always] be that of a proximate mechanism .... It might be that sentiments are a necessary condition for the existence of a society but, without the selective advantage, the explanation was not sufficient. For this reason, psychology could take but a limited role in the scientific explanation of social phenomena.

(Ghiselin, 1974, 218)

Ghiselin may be alluding here to a section in the third chapter of *The Descent of Man*, entitled "Moral Sense." Within this section is one of Darwin's most intriguing comments which bears directly upon the challenge that high intelligence poses to adaptationism: he acknowledges that various and often competing psychological mechanisms play a role in mediating social behavior in animals of higher intelligence, though the *origin* of social behavior must be independent of the evolution of significant cognitive complexity:

In many cases it is impossible to decide whether certain social instincts have been acquired through natural selection, or are the indirect result of other instincts and faculties, such as sympathy, reason, experience, and a tendency to imitation .... With respect to the impulse which leads certain animals ... to aid each other in many ways, we may infer that in most cases they are impelled by the same sense of satisfaction or pleasure which they experience in performing other instinctive actions; or by the same sense of dissatisfaction, as in cases of prevented instinctive action .... In many cases, however, it is probable that instincts are persistently followed from the mere force of inheritance, without the stimulus of either pleasure or pain.

(Darwin, 1871, 81–82)

Here we see Darwin wrestling with the challenge that complex emotional experience and high intelligence presents to those seeking to tease ultimate adaptive causes from a sea of proximate psychological mechanisms. Clearly Darwin, as well as Wilson and Ghiselin, did not believe that the higher faculties swamp or marginalize the innate mechanisms that descend to us through phylogenetic history. Although we may have difficulty imagining mechanisms by which non-hedonically or non-egoistically motivated behaviors arise in humans "through mere force of inheritance," we can all accept Darwin's implicit motivational pluralism in light of the cognitive modularity that modern neuroscience has established. Perhaps some variety of hedonism has always been appealing because it "seems to have good biological credentials," and introspection tells us that as drivers of behavior, pleasure and pain work pretty darn well (Sober and Wilson, 1998, 325). However, as Darwin's and Ghiselin's comments suggest, prosocial instincts often engage hedonistic or other "psychological" motivational factors but do not *necessarily* do so. And although pleasure (or other more complex *qualia* or cognitions like guilt for example) may contribute a motive vector to prosocial behavior in intelligent animals, the performance of many behaviors are quite persistently maintained despite having little or no hedonic salience or any purpose consistent with consciously formulatable goals or priorities. Furthermore, the existence of eusociality and altruism in ancient species implies that highly complex, prosocial behaviors were ancestrally available prior to more recently evolved complex emotional and cognitive processing.

Such arguments touch upon what Dennett has referred to as Darwin's "deeply counterintuitive 'inversion of reasoning'" (Dennett, 2009, 10061). As Dennett explains, one of Darwin's most vocal critics, Robert Mackenzie, was outraged by Darwin's apparently paradoxical thinking: namely, that complex natural phenomena (including behavior) could display utility and purpose without there being a manifest intelligence behind their design. Mackenzie had argued that "[Darwin], by a strange inversion of reasoning, seems to think Absolute Ignorance fully qualified to take the place of Absolute Wisdom in all of the achievements of creative skill" (Mackenzie, 1868). In other words, Darwin's critics—including Makenzie who coined the term "inversion of reasoning"—could not conceive of a process apart from something resembling human reason that could impart sense and purpose to things. A beaver may build a dam in complete ignorance of its design features (through "sheer force of inheritance"), but there must always be something resembling human reasoning operating—albeit displaced toward a metaphysical abstraction (i.e., a creator god, modeled after the thinking species, who made the beaver). Sense could never be derived from no-sense; there is never intelligibility without intellection! But, as Dennett explains, Darwin was able to neatly topple this metaphysical delusion by showing how massive complexity could be derived from blind and basic operations of the physical universe (Dennett, 2009, 10061).

The relevance of this to my indictment of the therapeutic narrative is hopefully obvious to the reader as many clinicians make the same metaphysical mistake Mackenzie makes: according to a narrative view of self, it is taken as a given that human behaviors always reflect intelligible human purposes, values, and goals; it is assumed that there is always an abstract-able sense to one's suffering and symptoms. Now, every therapist may recognize that compulsive hand-washing is an exception to this rule—but falling in love with the wrong person could never, in their eyes, have anything to do with fundamental laws of physics and chemistry (as instantiated in our DNA) or the remnants of ancient behavioral mechanisms.

None of this is intended to suggest that experience and mental constructs are trivial behavioral vectors. Of course in our species uniquely personal psychological motivations and cultural influences are all thrown into the mix along with base dimensions and mechanisms of behavior. But this is why the adaptationists (and sensible nativists) frequently warn about the pitfalls of confounding levels of analysis (the ultimate adaptive versus the proximate psychological). Appreciating the various levels that contribute vector force to human choice not only avoids theoretical missteps, it also helps explain why self-control is such a pervasive problem for *Homo sapiens*: in our unlucky species, goals, values, and desires are often at cross-purposes with the pre-specified parameters that have such an out-sized influence on behavior. As behaviorist Howard Rachlin explains, a squirrel always appears to "want" what is best (or most adaptive) for itself—it does not want to eat extra acorns in the fall but only to bury them (Rachlin, 2000, 16). Proximal and ultimate causes always coincide in the squirrel. In contrast, observe the human mother who actively dislikes her adult child but cannot stop fretting over that child's welfare, or the gambler who cherishes the family he destroys with each role of the dice, or the lover who is drawn to someone who represents an ongoing source of misery and chaos.

So then, Darwin's account of innate behaviors that appear to gain their motive force "directly through mere force of inheritance" suggests a way around the tangle of motivations, the myriad competing forces that so complicate explanations of behavior in humans. Darwin's "mere force of inheritance" argument has brought us around to a concept of innateness that can accommodate the problem of high intelligence—one that philosopher of science Richard Samuels has called the "psychological primitive." This refers to any behavior whose acquisition "cannot be explained by reference to any psychological process ... not just in this historical moment, but in principle" (Samuels, 2007, 25–26). These behaviors have no psychological antecedents, although their performance may be accompanied and influenced by emotive experience. The psychological primitive side-steps certain problems—such as trying to sort out which traits represent discrete adaptations or which traits are inborn vs. developed over time. But the concept trades these problems for others: for example, what criteria can be used to establish these primitives or distinguish them from closely derived, by-products or assemblages of true primitives? But despite such problems, the psychological primitive may be the best conceptual place holder for the weather-beaten concept of innateness. Commonsense suggests that discrete base dimensions of the behavioral phenotype must exist as more than useful abstractions: indeed, despite the complexity of the mammalian nervous system, the range of base properties undergirding behavior must be highly restricted in each species—delimited by the selective pressures that have shaped each species' emotional and motivational mechanisms.

The concept of the psychological primitive also jibes with recent interest in supplementing the current diagnostic schema with a dimensional approach to temperament and psychopathology—an approach that seeks to parse the behavioral phenotype into dimensional traits that show high levels of heritability and might eventually be linked to specific genetic or physiological markers. Candidate primitives include certain forms of impulsivity, affective instability, and risk aversion. One plausible primitive (or composite of other primary primitives) was curiously, but presciently, singled out by Wilson back in 1975—a trait he called indoctrinability. He theorized that this proclivity might have evolved from selective pressures that favored group cohesion and may underlie universal cultural innovations that foster such cohesion, such as religion (Wilson, 2000, 562). It also makes some persons vulnerable to political extremism and cult induction, or to positive responses to gently coercive therapeutic societies like Alcoholics Anonymous.

The psychological primitive is also reminiscent of Noam Chomsky's poverty of stimulus argument. Chomsky deduced that children could not acquire syntax by way of experience using domain non-specific (or so-called horizontal) cognitive processes like memory and abstraction. This appears irrefutable given the limited data available to young children and the rapidity with which they begin to master the vast complexity of linguistic communication. Therefore, Chomsky concluded, language must be a primary endowment—a "cognitive primitive" that children are born equipped to acquire. This brings us to a more recent tussle in the ongoing innateness wars: Chomsky versus Pinker.

Sixty years ago, Chomsky's theory of Universal Grammar sparked a resurgence of interest in innate mental structure and placed the precocious linguistics researcher at the vanguard of a nativist revival. Therefore it was perhaps not surprising that after Chomsky attended a working group in the 1970s to launch the left-wing attack on Wilson, he decided not to participate (Segerstrale, 2000, 205). In the ensuing years, Chomsky actually ventured into evolutionary speculation concerning the origins of the language faculty. However, this has been in tandem with, and in support of, his newer, highly recondite and controversial ideas on the nature of language known as the Minimalist Program (MP). Although many academics in linguistics have begun working within Chomsky's new framework, many in the field, including more traditional Darwinian thinkers like Steven Pinker, have been critical of MP. Some, such as Lappin and colleagues, have been downright incensed: they find MP to be more scientism than science—replete with inscrutable, unsubstantiated claims including the rather cryptic contention that grammar may be a "perfect" computational mechanism of optimal efficiency (Lappin et al., 2000, 666).

In one of his papers within this project, "The Faculty of Language: What Is It, Who Has It, and How Did It Evolve?," Chomsky and colleagues argue that the complex grammatical features of language, that he himself helped to codify, is a sort of virtual complexity—and that the central innovation specific to humans achieved by the linguistic organ is in fact limited to something called syntactic recursion (Hauser et al., 2002). Recursion accounts for the unbounded quality that language achieves as an unlimited number of fragments can be nested within a grammatical sentence (e.g., Bill suspects that his mistress knows that his wife heard about their affair). Chomsky's stripped-down theory on the features and origins of language suggests that the core feature of language may not have been forged gradually by natural selection as a communication-dedicated adaptation. Rather linguistic recursion may have arisen abruptly—in tandem with, or as a by-product of, other cognitive functions such as navigation or the number sense—and that language, despite displaying computational "optimality," is actually poorly suited for interpersonal communication and was not selected for this purpose. Pinker's rebuttal to this paper, published with Jackendoff, argues that the recursion-only theory of language "ignores the many aspects of Universal Grammar that are not recursive such as ... case, agreement, tense etc." (Pinker and Jackendoff, 2005, 201). The authors conclude that Chomsky's reasoning is "circular" in that he evokes a simplified theory of language in order to conclude that such a minimally complex function could not constitute the communication-dedicated adaptation it has generally been assumed to be. Chomsky's views are clearly at odds with Pinker and Jackendoff's more traditional emphasis on language's apparent design features that are suggestive of a vast degree of communicative and adaptive utility (Pinker and Jackendoff, 2005, 231).

Despite a tinge of incredulity that creeps into his refutation of Chomsky's MP, Pinker has generally maintained a respectful posture toward the master. However, in interviews subsequent to the original 2005 paper, Pinker has speculated about reasons for Chomsky's strikingly unorthodox views concerning the features and origins of language. Aside from Pinker's technical objections to MP, he has also questioned whether Chomsky's political philosophy may play a role in the development of these more recent ideas. Pinker has remarked, and I paraphrase, that as an anarchist Chomsky's vision of human nature is marked by idealism—an idealism that contrasts with what Pinker would likely regard as more realistic conceptions of human nature. Specifically, Pinker argues that Chomsky appears to emphasize a spontaneous human tendency toward "cooperation and creativity" as opposed to the more pragmatic aspects of behavior. This idealism, Pinker concludes, lends itself to a view of language that minimizes its practical and adaptive utility (Pinker, 2010). Indeed, Chomsky's ideas are explicitly anti-adaptationist—suggesting that language may have emerged abruptly, perhaps as an exaptation of other computational faculties, and that its original function may have been to serve as a means for achieving internal thought. When language is stripped of its utilitarian function, it is but a small step that Chomsky needs to take to conclude that the core features of language may not constitute a gradually evolved adaptation selected for interpersonal communication.

Obviously comments about how someone's politics might influence their scientific ideas are usually considered less than polite—especially as Chomsky has repeatedly stressed that his political and linguistic theories have no relation to each other. Ignoring this admonition, Pinker appears to be hinting that without some external frame (i.e., a political or philosophical context), Chomsky's rather obscure ideas concerning the origins of language might appear even more arbitrary and difficult to fathom. Nonetheless, this bagatelle constitutes a fairly civilized skirmish in the otherwise messy innateness wars that date all the way back to Mackenzie, Owen, Darwin, and Huxley. It underscores that the enigma of innate mental structure will continue to attract fascination as well as controversy—whether in psychology, philosophy, evolutionary biology, or linguistics. Furthermore, most of us—even the most gifted among us if Pinker's critique is accurate— cannot help but harbor personal and idiosyncratic notions of a reified human nature. These tend to reflect a range of biases—from a *love thy neighbor as thyself* vision of the species to the profound pessimism of a Zapffe or a Schopenhauer. Thus, debates over innateness will likely continue to be influenced by political and philosophical leanings with many on the far left emphasizing the malleability of behavior under the sway of social, cultural, and economic influences while certain conservatives embrace a more fixed, essentialist view of human nature—a view that can be seen as endorsing the alignment of social and economic structures with presumed species-typical tendencies. Of course both prejudices have their pitfalls.

After this review, the reader can hopefully appreciate why innateness has proved such a mercurial and contentious concept—why attempts to define it often appear partial, conditional, inexact, or all of the above. In the end, the following two aspects of innateness may represent its most useful features for psychiatry: high heritability and psychological "primitiveness." Heritability—the proportion of variation in a trait that is due to genetic differences among individuals in a population—avoids strict genetic determinism while providing a quantifiable measure of what may descend through the generations. While somewhat abstract and difficult to establish criteria for, the psychological primitive may be the best conceptual place-holder we have for innateness—which is probably why Chomsky used a variation of it (the poverty of stimulus argument) in his original theory on the faculty of language.

We can apply these two criteria to all sorts of traits that may represent innate dimensions (or second-order composites) of temperament. Frugality, for example, is unlikely to be genetically specified in any deterministic or strict sense (i.e., no one will ever find "cheapness" genes); however, this trait likely shows significant heritability. Furthermore, although there may be environmental contingencies that influence the degree of frugality ("we suspect that grandpa is cheap partly because he grew up during the depression"), in many cases it is likely that frugality constitutes a psychological primitive or at the least a closely derived by-product of primary primitives—like risk aversion and compulsivity. Thus, in most cases, its acquisition may be independent of psychological/biographically based factors. Perhaps some of the same genes that underly caching and hoarding behavior in animals, such as ravens, play some role in behaviors that humans display in relation to possessions and capital. However wildly speculative this may sound, it is probably less so than the majority of narrative explanations that abound in our therapeutic culture.

To summarize, in this chapter I have sought to provide a Darwinian perspective on contemporary discussions of nativism. The outmoded "nature versus nurture" controversy in psychology had an instructive precursor in the "instinctual versus learned" debate in evolutionary biology in the decades following the modern synthesis. This dichotomy proved to be equally specious as the dynamic features of genetic expression increasingly came to light. Developmentally, as well as evolutionarily, the expression of traits is of course meaningless without reference to environmental factors; and obviously how an organism responds to extrinsic conditions is a key factor in differential survival and reproduction. As Haldane and others theorized, variable behavioral responsiveness may account for a phenotype-driven model of adaptation wherein plastic responses could be targets and engines of evolutionary change. This could play a role in the evolution of species with more complex behavioral repertoires. But however complex behavior might become, the psychological primitives underlying these complex behaviors would not be submerged or swamped by a general expansion of the cognitive apparatus and its product—culture. Paradoxically, the most critical advance in human cognition—language—can, in many respects, be considered independent of more general or integrative cognitive functions involving reasoning and abstraction—*in spite of* the role language plays in extending such faculties. Indeed, despite the vast expansion of the cortex and its associated repertoire of skills and categories of knowledge, the language faculty evolved in a way that engages minimal conscious effort and reasoning—like navigation in birds or dam building in beavers. Therefore the language faculty must be considered, in the words of Waddington, completely assimilated to the genome. And so the crowning feature of human intellection is, in the traditional sense, wholly instinctual.

References

Alcock, J. (2009). *Animal behavior: an evolutionary approach.* Sunderland: Sinauer Associates.

Crispo, E. (2007). The Baldwin effect and genetic assimilation: revisiting two mechanisms of evolutionary change mediated by phenotypic plasticity. *Evolution,* 61, 11, 2469–2479.

Darwin, C. (1871). *The descent of man, and selection in relation to sex.* London: J. Murray.

Darwin, C. (1887). *The variation of animals and plants under domestication,* vol. 2. New York: D. Appleton.

Darwin, C. (1889). *The descent of man, and selection in relation to sex.* New York: D. Appleton, 230, quoted in Ghiselin, M. (1969).

Dawkins, R. (1982). *The extended phenotype.* New York: Oxford University Press.

De Beer, G. (1960). Darwin's notebooks on transmutation of species, part IV. *Bulletin of the British Museum (Natural History), Historical Series,* 2, 164, quoted in Ghiselin, M. (1969).

Dennett, D. (2009). Darwin's "strange inversion of reasoning". *Proceedings of the National Academy of Science*, 106, (Supplement 1) 10061–10065.

Ghalambor, C., McKay, J., Carroll, S., Reznick, D. (2007). Adaptive versus non-adaptive phenotypic plasticity and the potential for contemporary adaptation in new environments. *Functional Ecology*, 21, 3, 394–407.

Ghiselin, M. (1969). *The triumph of the Darwinian method*. Mineola: Dover Publications.

Ghiselin, M. (1974). *The economy of nature and the evolution of sex*. Berkeley: University of California Press.

Griffiths, P. (2002). What is innateness? *The Monist*, 85, 1, 70–85.

Griffiths, P. (2004). Instinct in the '50s: the British reception of Konrad Lorenz's theory of instinctual behavior. *Biology and Philosophy*, 19, 4, 609–631.

Hamilton, W.D. (1964). The genetical theory of social behavior I, II. *Journal of Theoretical Biology*, 7, 1–52.

Hauser, M.D., Chomsky, N., Fitch, W.T., (2002). The faculty of language: what is it, who has it, and how did it evolve? *Science*, 298, 2, 1569–1578.

Lappin, S., Levine, R.D., Johnson, D.E. (2000). The structure of unscientific revolutions. *Natural Language & Linguistic Theory*, 18, 3, 665–671.

Lehrer, J. (2012). Kin and kind. *The New Yorker*, 3, 5.

Lewontin, R. (1979). Sociobiology as an adaptationist program. *Behavioral Science*, 24, 1, 5–14.

Mackenzie, R. (1868). *The Darwinian theory of the transmutation of species examined*. London: Nisbet and Company.

Nowak, M.A., Tarnita, C.E., Wilson, E.O. (2010). The evolution of eusociality. *Nature*, 466, 26, 1057–1062.

Pinker, S. (2010). Interview, 4/21/2010: http://mitworld.mit.edu/video/160/

Pinker, S., Jackendoff, R., (2005). The faculty of language: what's special about it? *Cognition*, 95, 201–236.

Rachlin, H. (2000). *The science of self-control*. Cambridge: Harvard University Press.

Richards, R.J. (1987). *Darwin and the emergence of evolutionary theories of mind and behavior*. Chicago: The University of Chicago Press.

Ruse, M. Wilson, E.O. (1985). The evolution of ethics. *New Scientist*, 17, 52.

Samuels, R. (2007). Is innateness a confused concept? In Carruthers, P., Laurence, S., Stich, S., (Eds), *The innate mind (vol 3): foundations and the future*. New York: Oxford University Press.

Segerstrale, U. (2000). *Defenders of the truth: the battle for science in the sociobiology debate and beyond*. New York: Oxford University Press.

Simpson, T., Carruthers, P., Laurence, S., Stich, S. (2005). Nativism past and present. In Carruthers, P., Laurence, S., Stich, S., (Eds), *The innate mind: structure and contents*. New York: Oxford University Press.

Sober, E. (1993). *The nature of selection: evolutionary theory in philosophical focus*. Chicago: The University of Chicago Press.

Sober, E., Wilson, D.S. (1998). *Unto others: the evolution and psychology of unselfish behavior*. Cambridge: Harvard University Press.

Spurway, H. (1955). The causes of domestication: an attempt to integrate some ideas of Konrad Lorenz with evolutionary theory. *Journal of Genetics*, 5, 1, 325–362.

Westneat, D.E., Fox, C.W. (2010). *Evolutionary behavioral ecology*. New York: Oxford University Press.

Wilson, E.O. (2000) [1975]. *Sociobiology: the new synthesis*. Cambridge: The Belknap Press of Harvard University Press.

7 The Missing 50%

Non-Heritable Sources of Variance

This chapter attempts to gather up some of the loose strands that this polemic against narrative has tossed at the reader. Topics in the nativist/empiricist debate most relevant to psychiatry that have been broached in the above chapters will be more fully addressed: for example, what exactly accounts for non-heritable influences on behavior if the "shared" environment appears to exert so little influence? How can complex predilections and pathologies pass down the generations when behavioral traits appear to be controlled by such a large number of interacting factors?

As discussed above, most behavioral traits show a heritability of approximately 50%. This implies that half of the variance in these traits in a population can be attributed to genetic differences. Therefore the remaining 50% (which some have referred to as the "missing 50%") was by default considered non-heritable and thus generally assumed to be a function of environmental influence. However, as also mentioned in earlier chapters, adoption studies have also confirmed that the so-called shared environment (when not extreme) appears to exert small effects—we know this because the degree of concordance of mental traits in monozygotic twins does not vary significantly whether they are raised together or apart. Therefore many researchers had previously, by default, invoked the concept of a "non-shared" environment as supplying a significant portion of the non-heritable influence. This non-shared environment would include influences that are typically not shared by siblings living in the same household, such as peer groups, differential parenting, mentors—in short, all the experiences and relationships unique to the individual. However, many have questioned the cogency of this explanation: if sharing a home environment does not make monozygotic twins or adoptive siblings significantly more similar than if they were raised apart, why should non-shared experiences have any special potency—a

DOI: 10.4324/b23263-11

potency that shared experiences seem to lack (Mitchell, 2018, 85)? Indeed, a number of empiric studies of non-shared environmental factors (such as birth order and relationships outside the home) have failed to show consistent patterns of influence. It would appear that monozygotic twins will always show a significant and fairly stable degree of phenotypic divergence—irrespective of the degree of environmental congruence. Indeed, inbred mouse strains that are virtually isogenic and raised in standardized environments ("environmental twins") still show inter-individual variability that is in fact similar to that found in isogenic cohorts raised in variable, naturalistic settings (Wong et al., 2005).

But even if we allow for modest effects from non-shared experiences, how can we account for the lion's share of the missing 50%—the apparently non-heritable (hitherto presumed environmental) influence if traditional notions of the environment, either within or outside the home, do not appear to exert much in the way of systematic effects—at least that can be qualified or quantified at this time? What, then, are the sources of the significant behavioral discordance found in monozygotic twins? It seems that while population geneticists and social scientists were fighting over potential environmental sources of variance, molecular and developmental genetics came of age and discovered a "third component"—a source of variance derived neither from the environment in any deterministic or predictable manner nor from the actual base pairs that constitute our DNA, but rather derived from two main extra-genomic sources: the dynamic manner in which gene expression is regulated and the highly variable, unpredictable nature of embryonic brain development. Now as these processes are highly stochastic, it is easy to see how these sources of variance end up magnifying twin discordances and so inflate the non-heritable (presumed environmental) side of the tally, even though many of these processes may involve little in the way of extrinsic influences on the developing organism.

Perhaps the most important and most widely studied of these non-sequence-based sources of discordance in twins are known as epigenetic effects. These gene regulatory mechanisms, which can modulate phenotype without altering the actual DNA sequence, have become the subject of a burgeoning body of research. One of the most important of these mechanisms—DNA methylation—involves the attachment of a small molecule, called a methyl group, to certain bases within the DNA. Regions of DNA that are heavily methylated are not transcribed into gene products. Another type of epigenetic modification involves the physical packaging of DNA into chromatin—bundles of discrete stretches of DNA that are wound around proteins known as histones. The specific structural characteristics of a segment of chromatin (the types and density of proteins it contains, how tightly it is wrapped, etc.) will determine which regions of DNA will be accessible to be transcribed into gene products and which segments will be suppressed. Other important gene regulatory mechanisms which may act independently or in conjunction with the above mechanisms include genomic imprinting—wherein differential gene expression depends on whether a gene (and its transcriptional regulation) is inherited from the mother or the father. Additionally, there are microRNAs (miRNAs) which are non-coding fragments of RNA that target messenger RNA and down-regulate the translation of mRNA during protein synthesis. This is just a simplified and partial description of the molecular processes that control gene expression. These mechanisms work in tandem with and influence the activity of thousands of transcription factors—the proteins that directly initiate and control the transcribing of DNA into RNA.

In contrast to the actual DNA sequence, where mutations arise infrequently, the epigenetic profiles of cell lines in developing embryos are maintained with weak fidelity; additionally, *de novo* methylation patterns can arise in 3–5% of mitotic cell divisions. Therefore monozygotic twins may be discordant for many traits (and diseases) not because of environmental influences but because development is influenced by an epigenetic system that is only partially stable and therefore a significant source of variance that is random in nature (Wong et al., 2005).

A recent hypothesis concerning a possible epigenetic mechanism associated with homosexuality provides a good example of how this "third component" can influence the transmission of traits but in ways that often produce highly variable outcomes as well as fairly low heritabilities in most pedigrees. It is generally accepted by the scientific community that sexual orientation is an innate, biological endowment and homosexuality does run in families. Nonetheless the heritability has been estimated to be fairly modest—between 10–20% in a number of studies, despite the fact that environmental factors (in particular shared environmental influences) appear to exert little to no effect. As might be expected with these modest levels of heritability, no specific DNA polymorphisms have been associated with homosexuality. Thus the inheritance of this trait may never be linked to the actual DNA sequence but rather to other molecular mechanisms, like the epigenetic systems alluded to above, that are far from stable and often produce variable phenotypic outcomes in a scattershot manner. In fact, a team led by evolutionary biologist William Rice has identified a credible epigenetic mechanism that regulates fetal sensitivity to androgens during critical phases of embryological development and that may contribute to some forms of homosexuality (Rice et al., 2012). This is one example among many demonstrating that modest or even low concordance values for a trait in monozygotic twin studies (and the absence of associated DNA polymorphisms) do not in any way imply that the trait is influenced by environmental factors in any traditional sense.

To cite another example of how epigenetic mechanisms tend to owe more to the vagaries of molecular switches than anything resembling human logic, researchers at the University of Pennsylvania described a possible transgenerational epigenetic effect whereby cocaine use in male rats may actually *protect* their male progeny against the reinforcing effects of the drug. The putative mechanism involves the modification in a gene promoter region leading to the increased expression of a protein (brain-derived neurotrophic factor) in the prefrontal cortex that may mediate this behavioral brake on reinforced behavior (Pierce and Vassoler, 2014). So while I do not mean to imply that epigenetic regulation of gene expression is all noise with little signal, the outcomes do not necessarily follow anything resembling psychological sense.

Perhaps the largest and fastest-growing area of epigenetic research relevant to human psychopathology has focused on how early (including prenatal) stress can lead to short and long-term epigenetic modification of the hypothalamic–pituitary–adrenal (HPA) axis which constitutes the neuroendocrine region that forms the core of the mammalian stress response system. In numerous animal and human studies, epigenetic modulation of various genes that regulate the HPA axis (as well as the glucocorticoid receptors that are also critical in mediating stress) have been shown to be impacted by various environmental stressors. Such research into these crucial gene x environment interactions has begun to unravel specific mechanisms that likely play a role in triggering stress-related psychiatric disorders such as certain types of depression and anxiety (Klengel and Binder, 2015).

Another fertile area of research on the epigenetic effects of early stress or deprivation have focused on the ancient peptides vasopressin and oxytocin. These molecules have critical roles in mediating social behaviors in mammals such as monogamy and parental nurturance. In a paper entitled "Early Nurture Epigenetically Tunes the Oxytocin Receptor," the authors present data suggesting that reduced nurturance of prairie vole pups affects the methylation of genes that code for the oxytocin receptor. These epigenetic changes result in a reduction in oxytocin receptor density in critical brain regions which may lead to impairments in prosocial behavior after maturation (Perkebile et al., 2019).

As one might predict, the potential implications of this type of research for human psychological development have been eagerly anticipated by those in the narrative camp. Such findings are indeed helping to illuminate the dynamic interplay of extrinsic and genetic factors that characterizes development. Again, epigenetic effects are by no means all a function of random processes; nonetheless, many of these molecular notations and edits belong to the realm of biology and not the realm of human meanings and mental constructs. Where these modulators do conform to commonsense notions of disease triggers and susceptibilities, these likely resemble a cumulative-impact-of-stress model (that is most relevant to the HPA system) rather than acting as sculptors of the specific behavioral traits that constitute our core identities and proclivities. Even as more and more credible gene x environment mechanisms are identified, the ability to predict phenotypic outcomes will likely remain difficult. This is true, as one research group explains, because "phenotypes represent an emergent property of a dynamic biological system rather than the deterministic output of either genetic or environmental inputs" (Haque et al., 2009, 138).

As noted above, another significant source of variance that contributes to the discordance seen in monozygotic twins is supplied by the unique manner in which each embryonic brain develops. This process must of course be highly specified in order to produce the functionally and structurally distinct regions of the brain; nonetheless every developing brain follows a course that is fundamentally probabilistic and unique—like the fractal patterns of branching tree limbs. Development does not proceed according to a deterministic blueprint; rather it is influenced by all sorts of stochastic factors—the physicochemical properties of membranes and molecules, various gradient and positional effects, synapse formation, cellular migration, and so on (Mitchell, 2018, 69).

So then, even if molecular biology achieved the status of Laplace's Demon, one would still have difficulty predicting which monozygotic twin will develop schizophrenia and which will be spared. This is so because uncertainty in developmental genetics is not merely a reflection of our incomplete knowledge—rather uncertainty is intrinsic to a process that is fundamentally probabilistic. Indeed, beyond the most obvious source of indeterminacy (that introduced by mutation of the actual code), there are as well the accumulation of epigenetic marks, the induction of cellular differentiation through positional effects on embryonic cells, the sweeping planes of neuronal migration—all which suggest that blind luck may constitute a significant percentage of the missing 50% that makes each monozygotic twin unique.

Turning to that portion of the missing 50% that is actually tied to traditional notions of environmental influence, no one could deny that life experience (one's personal biography) may indeed be a potent force—especially if environmental variance is extreme. However, most of life's contingencies tend to yield highly constrained outputs according to one's enduring proclivities: a dozen different stressors will reliably provoke panic attacks in one person while the identical dozen stressors will provoke addictive craving in another—each according to their predisposition. No matter which contingencies arise, a panicked teetotaler would never think of pouring himself a drink when stressed, whereas the addict can think of nothing else. Externalities do not fundamentally shape us, rather they are permissive of our "natures." As previously mentioned, Darwin had presciently anticipated this truism:

> We are ... driven to conclude that in most cases the conditions of life play a subordinate part in causing any particular modification; like that which a spark plays, when a mass of combustibles bursts into flame—the nature of the flame depending on the combustible matter, and not on the spark.
>
> (Darwin, 1887, 282)

To update the analogy, the "combustible matter" is, of course, the genetic endowment and the "sparks" are the contingencies—both internal and external to the organism—that influence how this endowment is instantiated throughout development. A few examples can illustrate how environmental factors often exert their influence in ways that are both highly constrained (by genetic endowment) yet also unpredictable and unsystematic (i.e., not amenable to tidy narrative explanation).

The first example—a hypothetical one that I shall call "the fable of the fiddle and the guitar"—illustrates how seemingly trivial, chance, environmental factors can have outsized, unpredictable effects. Consider monozygotic twins that, for illustrative purposes, we can assume are also epigenetically and neuro-developmentally identical (let us pretend they are truly identical clonal beings). They are handsome, have high IQs, and musical talent. They are also personable, but somewhat passive and lacking in confidence. There is a family history of depression on their maternal side. On their tenth birthday, Twin A receives a guitar and Twin B a violin. Their parents are struck by the fact that both twins favor late Renaissance and early baroque music. While at conservatory, Twin A plays guitar in a blues band at a local dive to make money, while his brother plays violin in a local orchestra.

A troubled but beautiful woman seduces Twin A at the bar. She has a drinking problem and a personality disorder. Meanwhile, across town, Twin B is seduced by the second violinist at the orchestra. She is high-functioning and from a wealthy, well-adapted family. As mentioned, both twins are somewhat passive and tend to acquiesce in relationships. They both marry these women after graduating. Twin A has a child and his wife's behavior deteriorates; she consumes his savings with ineffective rehabs and she finally runs off with a local dentist. He becomes depressed and, burdened by loneliness and financial stress, hangs himself. Twin B, meanwhile, has a relatively stress-free life. His wife's trust fund enables him to pursue a career in music and he has well-adjusted children. He misses his brother and struggles with low-grade depression but is generally well. Convinced that the gift of the guitar killed his brother, he refuses to give his children music lessons. Are environmental contingencies important? Yes, terrifyingly so—but again, not according to any sort of neat narrative account. Rather, they are important in the inscrutable and implacable manner that religions and philosophies throughout the ages have attested to: God writes straight with crooked lines; it is Allah's will; fate guides those who will, those who won't she drags.

As discussed in Chapter 1, after decades of trying to find non-shared environmental factors that show any consistent or systematic influences on development, Robert Plomin, one of the world's preeminent behavioral geneticists, has adopted a similarly fatalistic conclusion concerning non-shared influences:

> the key environmental influences making us who we are might be down to chance, unpredictable events .... After thirty years of searching, it's time to accept the gloomy prospect. Non-shared environmental influences are ... idiosyncratic, serendipitous events.
>
> (Plomin, 2018, 80)

However dramatically experiential factors impact on our trajectories, the impact they have on the core proclivities that give our behavior and our choices their striking consistency throughout our lives is dwarfed, in Plomin's view, by genetic factors. Furthermore, as previously discussed, environmental factors are *themselves* highly influenced by heritable, constitutional factors (the so-called "nature of nurture"). Thus when the confounding effects of genetic differences are controlled for, the relatively small effect sizes of non-shared environmental factors are reduced even further (Turkheimer and Waldron, 2000, 90).

Despite this, geneticists such as Turkheimer continue to believe in the importance of studying environmental influences. He has argued that the shared/non-shared distinction was misleading—it matters less whether an event is objectively shared by siblings, but rather how events differentially impact persons with different temperaments. For Turkheimer, the project of finding non-shared environmental influences with significant effect sizes has largely failed for the same reason that the hunt for shared influences came up empty—or for that matter, why attempts to link genes with phenotypic traits are often difficult to replicate. This is true not because environmental factors are necessarily of minor importance; rather the dynamic and interactive aspects of development—each state being contingent both on temperament and on prior developmental states—ensures that each individual's developmental course will be idiosyncratic and highly indeterminate. Thus extrinsic factors that do make a difference will vary from person to person and so statistical measures of environmental factors in a population (that superimpose data from many individuals) are unlikely to identify influences that show consistent effect sizes of significant magnitude (Turkheimer, 2000).

But the gloomy prospect, according to Turkheimer, should not lead us to overly discount the import of non-genetic factors; rather the gloomy prospect appears to be more a function of the unpredictability and opacity of developmental processes and the limits of our experimental models and statistical tools. Furthermore genomic-based research is not immune from these same limitations (Turkheimer, 2000). But whether one is in Plomin's more nativist-leaning camp or in Turkheimer's, the gloomy prospect still stands and its implications for the narrative view of the self are the same: establishing systematic environmental sources of behavioral variance is extremely difficult; effect sizes of environmental influences are small and difficult to replicate. This returns us to the central argument of this monograph: *biographically based formulations of behavior are highly speculative; such formulations are often dominated by considerations of coherence, intelligibility, and ideology rather than plausibility.*

Turning to another instructive parable: consider a man with high genetic loading for addiction. He has lived his entire life in a strict Muslim country where alcohol is difficult to obtain. Although he displays some addictive proclivities (he plays the national lottery excessively), he has never had a drink. Apparent good fortune allows him to immigrate to Britain where he is offered a job working for a well-off relative. He's pleased with his new career and can afford to have his wife join him in the UK. However, within a year, the man develops a severe alcohol addiction and his new life is soon in shambles. An environmental factor (the availability of alcohol) functioned as a releaser of a genetic susceptibility. The reader might consider: What doomed this man? Was it his decision to immigrate or a few misplaced base pairs? How potent an environmental trigger would have to be for an individual to take this or that fork in the developmental landscape is highly variable—depending obviously on both the degree of genetic loading and the severity of the external trigger. For someone with a very significant genetic predisposition, the environmental trigger might be relatively small (in our example, the mere availability of an intoxicant); in cases of more modest genetic vulnerability, one might need the availability of the substance plus an elevated level of stress.

The phenomenon of indoctrination, that E. O. Wilson showed particular interest in back in the 1970s, exemplifies how experience often functions less as causative of pathologies or shapers of identities and more as releasers—vectors that push one down this or that groove in the developmental landscape. Indoctrinability is of course an abstraction; like many behavioral traits, it likely represents an amalgam of more basic, underlying, behavioral dimensions—such as suggestibility or submissiveness. Wilson actually believed the trait may have evolved to foster social cooperation and group cohesion. But no matter which underlying traits contribute to indoctrinability, one could reasonably expect these to be as heritable as most behavioral traits—that is, somewhere in the range of 40–50%. Therefore, a randomly selected population would present a bell-shaped distribution in terms of susceptibility to indoctrination. Now, unlike the above example of alcoholism, indoctrination often requires extreme and highly specific environmental conditions to fully emerge and so even those with a significant predisposition to indoctrination may never manifest the behavior. Nonetheless, susceptible persons would likely display related behaviors, throughout their lives, in the absence of frank indoctrination—behaviors such as strong placebo responses, a greater readiness to experience romantic infatuation, or a tendency to fall under the influence of a more dominant partner. This illustrates a paradoxical aspect of behavior: it is both highly responsive to near-term contingencies yet confined within the grooves of each individual's developmental course and quite stable and resistant to change over the long term. Another illustrative example of this truism comes from the Vietnam War. The widespread availability of cheap heroin and the stress of combat caused a large percentage of soldiers, including many with no family or personal history of addiction, to become hooked on the drug. But interestingly, when these soldiers were repatriated, an uncharacteristically high percentage of them were able to recover fairly rapidly with minimal treatment. This is something not usually seen in a more typical population of drug addicted individuals—many of whom are genetically predisposed to addiction.

As previously discussed, traits such as indoctrinability and addictiveness are, like most behavioral traits, artificial abstractions—second-order composites or by-products of more basic behavioral dimensions. Accordingly, such traits are likely influenced by scores of genes and genetic interactions. Given this, it is striking that (1) there is a fairly limited number of ways that behavioral disturbances manifest in our species *and* (2) that these complex amalgams still manage to pass so reliably through multi-generational pedigrees. But these two phenomena may actually be related: both reflect the fact that natural selection does not "select" or directly choose genes, rather selection operates on phenotypic traits, and these traits—even suites of traits—tend to enjoy a degree of stability. Hence, selection can forge behavioral repertoires in which complex traits and behaviors descend through the generations despite involving scores of interacting factors. One explanation for this remarkable feature of inheritance is that selection only generates a limited, highly-specified array of phenotypic outcomes due to species-specific developmental constraints operating during the formation of the embryonic brain. Generations of developmental geneticists since Waddington have been intrigued by this paradoxical aspect of development—that it is both unpredictable, involving all sorts of indeterminant processes, and yet it also appears buffered and constrained. This is so because various genetic regulatory mechanisms controlling development tend to specify a highly limited variety of phenotypic outcomes. Thus the developing organism toggles among pre-specified alternatives as it makes its way down the canalized developmental landscape that Waddington so presciently described decades ago (Mitchell, 2018, 76).

This remarkable property of development is known as robustness. It is characteristic of all species and though its adaptive function seems apparent—it prevents every small alteration or error in the genotype from impacting the phenotype—the mechanisms underlying robustness have been challenging to elucidate (Green et al., 2017). There are clearly some discrete, dedicated, molecular mechanisms such as the heat-shock chaperone proteins in Waddington's famous fly experiment that suppress alternative phenotypic expressions. But there are also global properties of gene expression that appear to underlie robustness across diverse cell and tissue types. The so-called "nonlinearity" that characterizes how genotypes map onto phenotypes might be one of the most important of these properties. Nonlinearity refers to the fact that fairly large changes in the genetic expression of a gene product may be tolerated during development with little or no impact on the target phenotype. However, once a certain threshold is reached, small changes in gene product then produce significant phenotypic effects. This nonlinear feature of gene to phenotype transformation may be one of the most important mechanisms responsible for developmental robustness and the canalization of traits that Waddington first described (Green et al., 2017).

Returning to behavior, the phenomenon of robustness may also help explain why—although most complex traits are derived from hundreds if not thousands of genes—a taxonomy of behavioral traits may be more than a heuristic: addictiveness, compulsivity, narcissism, and so on may pass so reliably through the generations (and represent common, stable features of the *Homo sapiens* behavioral repertoire) because these traits enjoy a degree of robustness—arising as they do from a limited range of developmental outcomes that have been canalized by species-specific developmental constraints *and* by the selective pressures that have sculpted suites of traits that form the core domains of the human behavioral phenotype. In a classic 1991 paper, evolutionary biologist Pere Alberch provided an experimental example that nicely demonstrates the paradoxical nature of development—being both reactive to extrinsic influences to a degree yet also constrained by the nature of developmental processes. In Alberch's experiment, frog embryos subjected to a developmental stressor (a mitotic inhibitor) adhered to their phylogenetic proclivity and always lost the first digit while salamander embryos subjected to the same stressor always dropped the fifth digit. Thus, as Alberch explains, "even if the parameters of the system are randomly perturbed ... the system will generate a limited and discrete subset of phenotypes" (Alberch, 1991, 7).

So then, the discrete variety of pathological traits likely results from the limited range of stable states of the developing nervous system as well as from the fact that selection has produced highly specified emotional mechanisms for controlling behavior—mechanisms shaped by species-specific survival and reproductive constraints. Variation in the parameters that control these mechanisms produce the range of responses characteristic of *Homo sapiens* related to threat-response, novelty-seeking, reproduction, social dominance, consumption, impulsivity, affiliative behavior, and so on.

Given that most behavioral traits and pathologies are second-order composites—often embodying emergent features of various interacting neurobiological circuits that are in turn controlled by scores of genes and gene regulatory mechanisms—it should therefore come as no surprise that linking traits to specific genes has proved quite difficult. Early on, candidate genes were discovered that appeared to correlate with discrete traits: for example, various polymorphisms related to serotonin transmission were linked to impulsivity and aggression as well as impaired stress tolerance and heightened risk for anxiety and depression; novelty-seeking and addictive proclivities were linked to genes for a variant dopamine receptor. However, many of these findings have been difficult to replicate consistently—confirming both the heterogeneity of many behavioral traits *and* that the relationship between genotype and phenotype is often indirect and vexingly complex. It is now assumed that genes rarely directly code for specific behavioral traits or conditions; instead genes code for morphological and functional features of nervous systems. Thus, genetic encoding in a narrow sense (i.e., genes that code for proteins with direct phenotypic effects) has been supplanted by concepts such as *developmental encoding* (Pigliucci, 2010). This concept encompasses not only the epigenetic regulation of genetic expression and interactions among networks of genes but also the developmental regulatory mechanisms mentioned above.

In light of the complexity of genotype to phenotype transformations, the search for direct correlations between discrete genes and traits may continue to be a disappointing endeavor. As a result, many behavioral geneticists have changed tactics—forsaking candidate gene approaches for population-based methods that estimate broad genetic correlations with psychological traits. Genome-wide association studies (GWAS) start with a large population that shares a phenotypic trait and then searches the entire genomes of these individuals for single nucleotide substitutions (SNPs) that might be associated with that trait. These substitutions or polymorphisms are single variant nucleotides (the letters that make up the genetic code) that may have no or a very minimal role in actually coding for the trait in question. However, when added together, these correlated polymorphisms can provide an estimate of genetic risk for manifesting the trait that they are statistically associated with.

In all likelihood, successful identification of genes associated with behavioral traits or disorders will require the deployment of a combination of methods: genome-wide searches, mechanism-informed pursuit of candidate genes and modulators of gene expression, as well as animal models involving homologous behaviors. Cataloging the progress being made by geneticists is beyond the scope of this book; besides, as noted, research into candidate genes is still so preliminary that any review would be of limited practical value in any clinical discussion of heritable human behavior. In the following chapter, then, I will rely on my experience as a clinician to argue that even in the absence of specific polymorphisms reliably linked to disorders, a dimensional approach to diagnosis, as well as research and treatment, often provides a more cogent framework for understanding psychopathology than biographically oriented perspectives.

References

Alberch, P. (1991). From genes to phenotype: dynamical systems and evolvability. *Genetica*, 84, 5–11.

Darwin, C. (1887). *The variation of animals and plants under domestication*, vol. 2. New York: D. Appleton.

Green, R.M., Fish, J.L., et al (2017). Developmental nonlinearity drives phenotypic robustness. *Nature Communications*, 8, 1–12.

Haque, F.N., Gottesman, I.I., Wong, A.H.C. (2009). Not really identical: epigenetic differences in monozygotic twins and implications for twin studies in psychiatry. *American Journal of Medical Genetics*, 151C, 136–141.

Klengel, T., Binder, E. (2015). Epigenetics of stress-related psychiatric disorders and gene x environment interactions. *Neuron*, 86, 1343–1357.

Mitchell, K. (2018). *Innate: how the wiring of our brains shapes who we are.* Princeton: Princeton University Press.

Perkebile, A.L., Carter, S.C., Wroblewski, K.L., et al. (2019). Early nurture epigenetically tunes the oxytocin receptor. *Psychoneuroendocrinology*, 99, 128–136.

Pierce, C., Vassoler, F.M. (2014). Reduced cocaine reinforcement in the male offspring of cocaine-experienced sires. *Neuropharmacology Reviews*, 39, 238.

Pigliucci, M. (2010). Genotype-phenotype mapping and the end of the 'genes as blueprint' metaphor. *Philosophical Transactions of the Royal Society, Biological Sciences*, 365, 557–566.

Plomin, R. (2018). *Blueprint: how DNA makes us who we are.* Cambridge: The MIT Press.

Rice, W.R., Frieberg, U., Gavrilets, S. (2012). Homosexuality as a consequence of epigenetically canalized sexual development. *The Quarterly Review of Biology*, 87, 4, 343–368.

Turkheimer, E. (2000). Three laws of behavior genetic and what they mean. *Current Directions in Psychological Science*, 9, 5, 160–164.

Turkheimer, E., Waldron, M. (2000). Nonshared environment: a theoretical, methodological and quantitative review. *Psychological Bulletin*, 126, 1, 78–108.

Wong, A.H.C., Gottesman, I.I., Petronis, A. (2005). Phenotypic differences in genetically identical organisms: the epigentic perspective. *Human Molecular Genetics*, 14, supplement 1, 11–18.

8   The Dimensional Approach

When a patient ends up in the office of a psychiatrist with nativist leanings, they may occasionally ask: If you believe, doctor, that my difficulties are so heavily influenced by heredity why bother discussing them? A tenable response is that understanding one's enduring proclivities can help identify the types of maladaptive behaviors and choices that one is prone to. Examples are provided that illustrate how to begin to think about human behavior in dimensional rather than biographical terms. What follows is not intended to be a thorough review of research on credible behavioral dimensions of the human behavioral phenotype; rather the aim is to provide a sample of some of the clusters of traits most relevant to clinical psychiatry and to demonstrate how research in behavioral genetics and ethology can help inform a dimensional approach to diagnosis and treatment.

The dimensional approach to behavior attempts to divide up the behavioral phenotype into categories that make biological sense. Some of these categories may never be traceable to discrete neurobiological systems as they cut across broad behavioral domains. Other dimensions may be more easily linked to modular, discrete functionalities and therefore may eventually be traceable to clusters of genes that are critical to the organization of these modules. To cite one such example, enormous progress has been made in parsing the neurobiological systems underlying addictive behavior. As discussed in Chapter 4 on Balzac, scientists have made real progress in understanding impulse dyscontrol, pathological seeking, and the compulsivity induced by abstinence and withdrawal. And these behavioral vulnerabilities have been linked to highly specific, well worked-out circuitry in the forebrain and frontal lobes.

On the other hand, much less is understood about, for example, the anatomy and physiology of introversion and extroversion in humans. Nonetheless, these traits, like addictiveness, show significant heritability.

DOI: 10.4324/b23263-12

Remarkably, genetic research into the social behavior of another species—the canine—has begun yielding potential clues about the biology and genetics of human sociality. Genome-wide association studies of the dog genome have located polymorphisms strongly associated with prosocial canine behaviors. It turns out these polymorphisms have been under intense selective pressure and are correlated with the modern dog's extreme motivation to pursue social contact and cooperation with humans—proclivities that are not found in the wolf and which have helped dogs achieve their cozy niche as the upright primate's best friend. Remarkably these polymorphisms are located very near to a region of the major genomic deletion that causes Williams Syndrome in humans—a genetic disorder marked by loss of stranger avoidance and pronounced hypersociality. Further investigation of candidate genes within this region has located a gene connected with amygdala circuits involved both in the processing of threatening social stimuli and in the reinforcing effects of sociality (Toth, 2019). Of course the pathological hypersociality characteristic of Williams syndrome may involve different genes and mechanisms from those underlying run-of-the-mill extroversion. However, both the aberrant hypersociality seen in Williams Syndrome and the exuberant gregariousness found in the canine species demonstrate that highly complex social behaviors may indeed be traceable to discrete genetic variation. The leap to pedestrian human extroversion and shyness may not, in the end, be as far as one might imagine.

Returning to the behavioral dimension of impulsivity discussed at length in Chapter 4 on Balzac, this proclivity can be best understood as a tendency to make choices based upon immediate or near-term contingencies rather than on intermediate or long-term ones. A related feature of impulsivity appears to be a greater tendency to discount or ignore risk. Interestingly, these two dimensions of impulsivity are likely dissociable to an extent. For example, there are a subset of impulsive people who are also anxious and highly sensitive to risk. Obviously this tendency toward anxiety would likely have a moderating effect on their difficulties with temporally related impulse control.

Understanding the behavioral dimension of impulsivity allows one to appreciate a number of potentially related behaviors. Consider, for example, the proclivity toward excessive procrastination. Upon first consideration, this tendency might appear primarily related to problems with motivation. However, research suggests procrastination is actually strongly associated with impulsivity; thus, procrastination can best be understood as an impulsive response to distress—avoidance being the quickest means of squelching mental discomfort. Therefore, impulsive people may not only display classic impulse dyscontrol—like excessive spending or eating—but they may also engage in a range of behaviors that will most rapidly rid them of negative mental states. Hence, avoidance and procrastination, online shopping and pathological eating, are all related behaviors found in sub-groups of impulsive people. This may also explain why many people with bulimia (impulsive eating in response to stress) are often prone to excessive interpersonal accommodation (reflexively capitulating to the expectations of others): such behavior likely relieves near-term discomfort, albeit at the expense of longer-term personal needs or goals. Thus, knowing that a patient is prone to binge eating alerts the astute clinician that the patient may also likely be a procrastinator and an interpersonal accommodator. This is a good example of how dimensional thinking can help explain clusters of seemingly unrelated behaviors that may actually share common genetic underpinnings.

Reflexive interpersonal accommodation is not only associated with impulsivity, it is also linked to other heritable traits—specifically, elevated levels of so-called reward dependence (high sensitivity to signals of social approval and support) as well as higher degrees of empathic responsiveness. While some people alienate others with selfish, entitled behavior, others appear to have exaggerated empathic responses—one could call them "pathological empaths"—much like Adeline, the Baron's excessively forgiving wife in Balzac's novel. Like any other trait, the capacity for empathy in a population is distributed on a bell-shaped curve. People with very high levels often make devoted friends but can be tyrannized by excessive compunction or easily manipulated by guilt.

In contrast to the natural empath are those persons with very low levels of empathy. Some of these individuals appear to lack a visceral response to the suffering or misfortune of others. This deficit can be associated with severe forms of narcissism as well as the transgressive cruelty and criminality seen in antisocial personality disorder. Interestingly, animal models of empathic-like behaviors have been identified, suggesting that certain aspects of empathy may not require complex mental states, such as awareness of self and other. Indeed significant experimental evidence indicates that empathy is an evolutionarily conserved mechanism, present in other social mammals such as rodents. A recent review article, "Observational Fear Behavior in Rodents as a Model for Empathy," details various studies that strongly suggest mice and rats display affective sensitivity to social partners (Kim and Keum, 2019). Rodents have long been known to display prosocial proclivities such as consolation, and in the studies reviewed in this paper, a phenomenon known to be specifically correlated with trait measures of empathy in humans—observational fear conditioning—was demonstrated in a number of experimental designs. The data confirm that exposing an animal to another animal of the same species receiving aversive foot shocks will induce vicarious fear conditioning in the observer. The authors conclude that this social transmission of fear is a bona fide manifestation of affective empathy (Kim and Keum, 2019). As one would expect with true empathic phenomena, the degree of the vicarious response is influenced both by the degree of kinship and the level of social familiarity. The capacity for empathy, then, although of course influenced by all sorts of higher-level cognitive concomitants in *Homo sapiens*, is likely a base dimension of temperament—a psychological primitive. The relative degree of empathic responsiveness constitutes a particularly important dimension of the human phenotype, variations of which can be seen in a number of pathological types. Low levels can lead to lives marked by isolation and interpersonal failure while excessive levels can be equally crippling—associated in some persons with heightened reward-dependence (reliance on external inputs for esteem regulation) and leaving some vulnerable to interpersonal exploitation.

Sensitivity to threat or risk is another important dimension of behavior that is highly heritable and significant variation in this trait can be found in a number of disorders. Those with antisocial tendencies, for example, not only have reduced empathic responses, but also display a range of related emotional deficits (sometimes referred to as "callous-unemotional traits"). These can include globally reduced fear levels to potential threats to themselves as well as diminished emotional responsiveness to the distress of others (Szabo et al., 2019). Imaging studies have pointed to reduced activity not only in the amygdala (the center of fear processing) but also a region called the anterior cingulate gyrus. This area appears critical to the emotion neuroscientists have labeled "moral fear"—the emotional response critical for influencing behaviors that might carry repercussions for others. It is theorized that sociopaths certainly *know* right from wrong but they cannot *feel* the emotional discomfort that doing wrong arouses in normal individuals. This makes some of them as smooth as polished stone—even able to fool a polygraph test. It also suggests that cognitive experiences stripped of affective valence are much less likely to influence behavior.

At the other end of the spectrum from the sociopath are those patients with certain types of anxiety disorders. Some who suffer from obsessive compulsive disorder (OCD), for example, can actually resemble the mirror image of sociopaths in that they display excessive scrupulousness—an exaggerated concern over the consequences of their actions. In fact, a common symptom of OCD is the worry that one may have committed an unwitting crime or transgression. For example, a patient might develop the intrusive fear that they accidentally struck a fallen pedestrian on a dark road after going over a pot hole. Regions of the brain that are responsible for detecting and responding to threat and other categories of contingency appear to be overactive in these patients. This can yield a mindset dominated by the anticipation of all manner of negative outcomes as well as a surfeit of moral fear.

Obviously the innate traits that govern our behavioral responses are often the product of numerous subroutines and proclivities in varying degrees and combinations. For example, although we tend to think of impulse-ridden types as oblivious to risk, there can be—as mentioned above—impulsive persons who are also risk-averse and anxious. This may serve to temper their impulsive tendencies—at least in situations where risk appears more salient. The ways in which various traits combine and interact in individuals and pedigrees also helps explain how combinations of traits can lead to second-order correlations with highly non-specific social phenomena and behaviors that, by their very nature, one would never guess might run in families. Divorce is one such example of how a cultural invention—linked to highly variable behaviors—may be surprisingly correlated with genetic factors. Although many marital therapists assume divorce is chiefly passed down the generations through experience, it has in fact been shown to be more influenced by heritable rather than environmental factors—adoptees carrying the larger share of risk-of-divorce from their biological relatives not their adoptive families (Salvatore et al., 2018). This is true in a population sense only: there are, of course, no specific traits that could predict the risk of divorce with any certainty in an individual; rather one may inherit a constellation of traits that serves to elevate one's risk of divorce. Obviously the types of traits associated with something as non-specific as elevated-risk-of-divorce will vary tremendously from pedigree to pedigree. One could guess that in some families the elevated risk might be tied to traits that influence sexual infidelity—such as novelty-seeking and impulsivity. In other families, however, the risk of divorce might be tied to elevations in a range of psychopathologies that are relatively independent of mating behavior but which can interfere with lasting attachment and/or stable interpersonal function.

Narcissistic personality disorder (NPD) represents another example of how a highly complex amalgam of traits can still run in families to a remarkable degree. This disorder involves broad and fundamental aspects of a person's identity and ways of perceiving themselves and others. Given this, one would not expect its heritability to approach that of disorders characterized by classical symptom patterns involving more discrete neurobiological systems. Yet one recent Norwegian twin study reported a heritability for this personality disorder of 0.70— matching or exceeding that of disorders like bipolar that most would agree represent true disorders (Torgersen et al., 2012). Other studies indicate heritabilities of all Cluster B personality disorders (narcissistic, borderline, histrionic, and antisocial) in the 0.50 range which is still quite high. Of note, all studies indicate a zero or close to zero influence of shared environment. Thus, although it can be argued that NPD is an artificial construct, it appears to capture a constellation of traits that track together in certain pedigrees and so the disorder (at least in some families) does appear to represent some sort of natural kind and not simply a bundle of non-specific tendencies. Indeed, geneticists have established significant heritabilities for many of the core features of the disorder—like grandiosity, entitlement, and diminished empathy.

The dimensional approach can not only help illuminate the distinguishing features of certain natural kinds, it may also help explain certain phenomenal convergences and divergences such as why different personality types may display similar behaviors under certain conditions. For example, both those with narcissistic and borderline personality disorders tend to display transgressive interpersonal behavior (i.e., vindictive rage) that can disrupt social bonds. Nonetheless many persons with borderline personality appear capable, in a number of contexts, of significant empathic responsiveness. Perhaps in many instances, their excessive emotionality and impulsive reactivity so profoundly impair self-control that empathic or other sources of restraint yield no traction. In contrast, the transgressions of the narcissist may be more a function of a congenital lack of empathic capacity and/or a need to assert interpersonal dominance.

The dimensional perspective, informed and inspired as it is by ethology and evolutionary biology, is also always on the lookout for potentially relevant animal analogs of core human behavioral traits. For example, one of the more intriguing aspects of NPD is the paradoxical co-existence, found in many patients, of poor esteem-regulation together with high levels of interpersonal competitiveness/aggressiveness which often gives them a leg up in primate politics—whether in the office or on the street. One could speculate that genes and mechanisms related to social dominance in other animals might one day prove relevant to the study of the genetics and neurobiology of some sub-types of NPD, just as canine behaviors discussed above have shed light on some forms of hypersociality in humans.

As previously mentioned, candidate gene studies of human behavioral traits have proved difficult to replicate. However, research into the genetics of prosocial behaviors in other species—such as the prairie vole—has begun to unravel the genetic basis of critical social behaviors, such as those involved in pair-bonding. While the vast majority of avian species exhibit sexual monogamy, only 3–5% of mammals are considered monogamous, and of these only a tiny percentage display strict sexual monogamy. Therefore sexual fidelity is no longer considered the defining feature of protracted pair-bonding and so ethologists now use the term "social monogamy" to describe the affiliative behavior expressed by consorts that remain together after mating. Features of social monogamy include the defense of common resources, mate-guarding, cooperative breeding, and incest avoidance (Carter and Perkeybile, 2018).

Various forms of social monogamy appear to have evolved independently in many geographically isolated mammalian species and remarkably these behaviors are, without exception, mediated by two ancient peptides—oxytocin and vasopressin. As would be expected, there is inter- and intra-species variation in the location and density of receptors to these peptides (as well as how these receptors are influenced by androgens, estrogen, and stress hormones). These physiological variations (that have been traced to various genetic polymorphisms) predict variability in core monogamous behaviors such as aggression toward same sex animals after mating, levels of parental investment, as well as degree of sexual monogamy. Interestingly, oxytocin receptor density in a particular anatomical region—the nucleus accumbens—appears critical to the types and degrees of sociality observed. As discussed in Chapter 4 on Balzac and sex addiction, the nucleus accumbens is the brain region critical for the mediation of all reinforced behaviors—including those behaviors entrained by drugs and so-called natural rewards. This may explain why some prosocial and affiliative behaviors, romantic infatuation for example, are so compelling for humans. It may also explain why various addiction treatments employ forms of social affiliation (in the form of peer fellowship and sponsorship) with the hope that social reinforcers can help replace reliance on substance use. Interestingly, a number of studies have shown that isolated rodents will do considerable work simply to attain the reward of mere visual contact with a conspecific residing in a separate cage. For highly social mammals, then, mere inter-animal proximity may constitute an important category of reward!

There have been some promising leads on the genetics of prosocial behaviors in humans. One review paper identified five polymorphic genes that create variation in oxytocin/vasopressin pathways—including variations in the hormones, their receptors, and substances that influence their release and degradation (Epstein et al., 2012). For example, one study showed that normal subjects with a polymorphism in the gene coding for the oxytocin receptor displayed lower levels of empathy and affective reactivity. Studies in socially normal individuals have been difficult to replicate—possibly due to the broad behavioral and genetic heterogeneity within normal populations. However, a number of polymorphisms have been identified in cohorts with autism spectrum disorders. These genetic differences may provide clues to the genetic basis of these disorders as well as insights into the variation found in affiliative behaviors in neuro-typical populations.

As discussed above, findings regarding genetic influences on human behavior often show significance only at the level of populations. Thus, determinations about the influence of genes on traits in individuals may remain limited for decades to come. Nonetheless, thinking in terms of innate dimensional categories can yield valuable insights into the uncanny consistency displayed by human types. Thus, even if one cannot yet imagine a day when the father of the bride would check the prospective groom's cheek swab for polymorphisms indicating the future risk of philandering, I do advise patients to adopt a dimensional perspective when assessing traits in potential long-term mates that are relevant to pair-bonding behavior (as well as general psychological fitness).

As argued throughout this volume, thinking in terms of dimensional traits allows clinicians, researchers, as well as patients to reflect upon the ineluctable patterns of our choices and behaviors—the residua of which create each person's unique trajectory. This is fundamentally a descriptive, taxonomic approach—one that often eschews questions of ultimate causation. Obviously helping patients understand proximate drivers of decisions, emotional reactions, and behaviors can be very helpful for patients attempting to navigate their worlds, and this is always an important part of psychotherapy. However, as also argued above, establishing the ultimate causes of enduring patterns of behavior may be as pointless as it is speculative. More importantly, positing theories on the origins of behavioral proclivities may have limited effects on symptom remediation or future patterns of choice. So then, a treatment based on the dimensional perspective often opts for trading "why" questions for "how" questions: namely, how do patterns of choice operate under variable conditions? How can identifying default tendencies help one utilize reason and self-control to interrupt maladaptive patterns of behavior?

This chapter has obviously only skimmed the surface of the literature on the genetic influences on human behavior. I invite the reader to delve into the burgeoning scientific literature on genes, the brain, and behavior. One may read studies on the heritability of various traits, theories of credible animal models of human proclivities, and research into plausible mechanisms involving potential candidate polymorphisms. Much of this research is provisional but it is fascinating nonetheless—representing the first tentative steps in the search for the true psychological primitives of our species.

References

Carter, C. S., Perkeybile, A. M. (2018). The monogamy paradox: what does love and sex have to do with it? *Frontiers in Ecology and Evolution*, 6, 1–20.

Epstein, R. P., Knafo, A., Mankuta, D., Chew, S.H., Lai, P.S. (2012). The contributions of oxytocin and vasopressin pathway genes to human behavior. *Hormones and Behavior*, 61, 359–379.

Kim, A., Keum, S. (2019). Observational fear behavior in rodents as a model for empathy. *Genes, Brain, and Behavior*, 18, 1, e:12521.

Salvatore, J.E., Lonn, S.L., Sundquist, J., Sundquist, K., Kendler, K.S. (2018). Genetics, the rearing environment and the intergenerational transmission of divorce: a Swedish national adoption study. *Psychological Science*, 29, 3, 370–378.

Szabo, E., Halász, J., Morgan, A., Demetrovics, Z., Kökönyei, G. (2019). Callous-unemotional traits and the attentional bias towards emotional stimuli: testing their moderating role of emotional and behavioral problems among high-risk adolescents. *Clinical Child Psychology and Psychiatry*, 25, 1, 156–173.

Torgersen, S., Myers, J., Reichborn-Kjennerud, T., Roysamb, E., Kubarych, T., Kendler, K. (2012). The heritability of cluster B personality disorders assessed both by personal interview and questionnaire. *Journal of Personality Disorders*, 26, 6, 848–866.

Toth, M. (2019). Hypersociality: the other side of the coin. *Genes, Brain, and Behavior*, 18, 1–11.

Index